Philosophy of Biology: A Very Short Introduction

VERY SHORT INTRODUCTIONS are for anyone wanting a stimulating and accessible way into a new subject. They are written by experts, and have been translated into more than 45 different languages.

The series began in 1995, and now covers a wide variety of topics in every discipline. The VSI library currently contains over 600 volumes—a Very Short Introduction to everything from Psychology and Philosophy of Science to American History and Relativity—and continues to grow in every subject area.

Very Short Introductions available now:

ABOLITIONISM Richard S. Newman
ACCOUNTING Christopher Nobes
ADAM SMITH Christopher J. Berry
ADOLESCENCE Peter K. Smith
ADVERTISING Winston Fletcher
AESTHETICS Bence Nanay
AFRICAN AMERICAN RELIGION
    Eddie S. Glaude Jr
AFRICAN HISTORY John Parker
    and Richard Rathbone
AFRICAN POLITICS Ian Taylor
AFRICAN RELIGIONS
    Jacob K. Olupona
AGEING Nancy A. Pachana
AGNOSTICISM Robin Le Poidevin
AGRICULTURE
    Paul Brassley and Richard Soffe
ALEXANDER THE GREAT
    Hugh Bowden
ALGEBRA Peter M. Higgins
AMERICAN CULTURAL
    HISTORY Eric Avila
AMERICAN FOREIGN
    RELATIONS Andrew Preston
AMERICAN HISTORY
    Paul S. Boyer
AMERICAN IMMIGRATION
    David A. Gerber
AMERICAN LEGAL HISTORY
    G. Edward White
AMERICAN NAVAL
    HISTORY Craig L. Symonds
AMERICAN POLITICAL
    HISTORY Donald Critchlow

AMERICAN POLITICAL PARTIES
    AND ELECTIONS L. Sandy Maisel
AMERICAN POLITICS
    Richard M. Valelly
THE AMERICAN PRESIDENCY
    Charles O. Jones
THE AMERICAN REVOLUTION
    Robert J. Allison
AMERICAN SLAVERY
    Heather Andrea Williams
THE AMERICAN WEST Stephen Aron
AMERICAN WOMEN'S
    HISTORY Susan Ware
ANAESTHESIA Aidan O'Donnell
ANALYTIC PHILOSOPHY
    Michael Beaney
ANARCHISM Colin Ward
ANCIENT ASSYRIA Karen Radner
ANCIENT EGYPT Ian Shaw
ANCIENT EGYPTIAN ART AND
    ARCHITECTURE Christina Riggs
ANCIENT GREECE Paul Cartledge
THE ANCIENT NEAR EAST
    Amanda H. Podany
ANCIENT PHILOSOPHY Julia Annas
ANCIENT WARFARE
    Harry Sidebottom
ANGELS David Albert Jones
ANGLICANISM Mark Chapman
THE ANGLO-SAXON AGE John Blair
ANIMAL BEHAVIOUR
    Tristram D. Wyatt
THE ANIMAL KINGDOM
    Peter Holland

Available soon:

For more information visit our website

www.oup.com/vsi/

Samir Okasha

PHILOSOPHY OF BIOLOGY

A Very Short Introduction

OXFORD
UNIVERSITY PRESS

OXFORD

UNIVERSITY PRESS

Great Clarendon Street, Oxford, OX2 6DP,
United Kingdom

Oxford University Press is a department of the University of Oxford.
It furthers the University's objective of excellence in research, scholarship,
and education by publishing worldwide. Oxford is a registered trade mark of
Oxford University Press in the UK and in certain other countries

First edition published in 2019

Impression: 2

Published in the United States of America by Oxford University Press
198 Madison Avenue, New York, NY 10016, United States of America

British Library Cataloguing in Publication Data
Data available

Library of Congress Control Number: 2019946779

ISBN 978–0–19–880699–8

Printed in Great Britain by
Ashford Colour Press Ltd, Gosport, Hampshire

For my parents, and for Leila, Mona and Rami

Contents

Acknowledgements

I am grateful to the many generations of students at the University of Bristol to whom I have taught this material. For their comments on the manuscript, I thank Paul Griffiths, Andrew Howard, Philip Kitcher, Arsham Nejad Kourki, Walter Veit, and Lucas Wakling.

List of illustrations

Chapter 1
Why philosophy of biology?

Philosophy has long been intertwined with the natural sciences.
Many of the greatest philosophers of the last 400 years, including
Immanuel Kant and David Hume, were influenced by the science
of their day; and some, such as René Descartes and Gottfried
Leibniz, made important scientific contributions of their own.
This intertwining is not surprising. Since antiquity, philosophy
has asked questions about the nature of the universe, the place of
humans in it, and our knowledge of it; and these are matters on
which science also has much to say. Indeed, with the emergence of
modern science in the 17th and 18th centuries, many questions
that were traditionally the province of philosophers had to be
surrendered to the scientists. Examples include whether all matter
is made up of atoms, as the ancient philosopher Democritus held,
and whether the human mind is composed of a non-physical
substance, as Descartes held. No-one could sensibly discuss
these questions today without attending to what science says
about them.

Though the birth of science led to the colonization of parts of
philosophy, it also gave rise to a new type of philosophical enquiry,
which asks questions about the methods of science itself. Can
experiments ever prove that a scientific theory is true? Can all
scientific knowledge be reduced to a few fundamental principles?
If two scientific hypotheses both fit the data, can we rationally

choose between them? These are not scientific questions per se, but rather philosophical questions *about* science. That does not make them the exclusive preserve of philosophers. Indeed, scientists including Newton and Einstein have thought deeply about such questions. But in the early years of the 20th century, the study of the scientific method crystallized into an academic discipline in its own right, known as philosophy of science, which flourishes today. Contemporary practitioners typically have a training in both philosophy and science, and in some cases straddle them.

Philosophy of biology is a sub-branch of the philosophy of science that emerged in the 1970s and has grown rapidly since. In retrospect we can discern three factors behind its emergence. First, it become clear that traditional philosophy of science was too physics-centric—biology had been left out of the picture. Second, conceptual issues that arise within biology began to attract the interest of philosophers, leading to fruitful interdisciplinary exchanges. Third, proponents of 'naturalized' philosophy, which uses empirical science to help tackle philosophical problems, increasingly looked to biology for inspiration. These three factors correspond to the three main sorts of enquiry within contemporary philosophy of biology, so are worth expanding on.

In the early to mid-20th century, the dominant school in philosophy of science was *logical empiricism*. The original logical empiricists included Rudolf Carnap, Hans Reichenbach, and Carl Hempel, who emigrated from Europe to the US in the inter-war period. They had a background in physics, and the picture of science they developed took physics as its model. The logical empiricists emphasized 'laws of nature'—fundamental theoretical principles which underlie observed phenomena. They characterized scientific enquiry as the search for such laws, and scientific explanation as the deduction of phenomena from the laws. This picture applies quite well to physics, where we find an abundance of laws, such as the law of universal gravitation; and an abundance of phenomena, such as planetary motion, which can be deduced from them. But it applies less well to the biological sciences. If you open a textbook in any branch of biology, such as genetics, molecular biology, or zoology, you will find a wealth of empirical facts and a variety of models and theories that are used to explain them. However, you will be hard pushed to find any 'laws of nature' from which those facts can be deduced; that is simply not how the scientific information is organized. In this respect and others, the logical empiricist conception of science could not easily accommodate biology.

It was not just the logical empiricists who marginalized biology. In 1962, the philosopher and historian Thomas Kuhn published *The Structure of Scientific Revolutions*, a work that was critical of logical empiricism and contributed to its eventual demise. Kuhn focused on how scientific ideas change over time. He argued that a mature scientific discipline is organized around a 'paradigm', which is a set of shared assumptions that guide day-to-day research. During a period of 'normal science', the paradigm is not called into question but rather accepted unquestioningly, serving as the backdrop to enquiry. But occasionally a period of crisis occurs, culminating in revolution: the reigning paradigm is overthrown by a new one. Kuhn argued that such paradigm shifts are not driven by entirely rational factors, and that the new paradigm is not necessarily 'better' than the one it replaces, just different. Thus the history of science is not simply a linear march towards the 'objective truth', on Kuhn's view.

Kuhn's work was radical, challenging the received wisdom in philosophy of science. But like the tradition that he challenged, Kuhn's focus was on the physical sciences. His examples of paradigm shifts included the Copernican revolution, in which the geocentric model of the solar system was replaced with the heliocentric one; the Einsteinian revolution, in which classical mechanics was replaced with relativity theory; and the chemical revolution, in which the phlogiston theory of combustion was replaced with the oxygen theory. But Kuhn had little to say about the biosciences, despite writing in the heyday of the molecular revolution in biology. Nor did he discuss such obvious cases as the rise of Darwin's theory of evolution, the germ theory of disease, and the Mendelian theory of genetics. Like the logical empiricists, Kuhn's philosophy of science relegated biology to the sidelines. The philosophy of biology arose partly out of a perceived need to redress this imbalance.

The second motivation behind philosophy of biology was internal to science, rather than coming from philosophy. The 20th century witnessed tremendous advances in the biosciences. In the first half of the century evolutionary biology took centre-stage, as the 'neo-Darwinian synthesis' emerged. This involved the integration of Darwin's theory of evolution, suitably updated, with other biological disciplines such as genetics, palaeontology, and zoology. The result was an impressive theoretical edifice, based on the idea that evolution is driven by the natural selection of some genetic variants over others. In the second half of the century the emphasis shifted to molecular biology, which studies the molecular basis of living organisms and their cellular components. This enterprise began in the 1930s with experimental work on bacteriophages (viruses that attack bacteria). It received a major impetus in 1953 when James Watson and Francis Crick discovered the structure of DNA, the material of which genes are made. In the ensuing decades molecular biology burgeoned, as biologists unravelled the complexities of how genes work, and how they are copied. This in turn gave birth to modern genomics, which studies how an organism's full complement of genes affects its growth and development.

As philosophers became acquainted with these scientific developments, they discovered a wealth of interesting conceptual issues, ripe for philosophical analysis. To take one example, evolutionary biologists often employ a purposive idiom when discussing organisms and their traits. They talk about the 'function' of particular traits, or what they are 'for'. Thus a fish's gills are for breathing, and a cactus's spines are for deterring herbivores. Such talk is puzzling at first sight. In other fields of science, such as physics or chemistry, there is no comparable talk of function. So why do biologists talk this way, and what does it mean? Is it a hangover from a pre-Darwinian era when living organisms were thought to have been created by God? How do we discover what a trait's function is, and can it have more than one? This is a classic topic in the philosophy of biology, discussed in Chapter 3.

A second example comes from genetics. In this field one finds frequent use of concepts borrowed from human communication systems, such as 'coding', 'information', and 'translation'. Thus genes are said to encode information that is passed from parent to offspring, and that is 'read' during development, enabling a zygote to develop into an adult in the manner appropriate to its species. Again, such talk is rather puzzling at first sight. After all, a gene is ultimately just a special sort of macromolecule, made of DNA. We do not normally think of molecules as carriers of information, nor do we describe most molecular interactions in terms of information transfer. So why does informational language permeate genetics, and what does it mean? Should it be taken literally, or is it just metaphor? This topic is discussed in Chapter 6.

Again, it is not that these questions do not occur to biologists. On the contrary, the two examples above have been discussed by eminent biologists, including Jacques Monod, Ernst Mayr, and John Maynard Smith. But for the most part, practising biologists are more concerned with making empirical discoveries than with engaging in philosophical analysis of what their concepts mean. This is as it should be. Though interaction between philosophers and scientists is beneficial, there needs to be a certain distance between scientific practice and philosophical reflection. This allows the scientists to focus on their day job, and allows philosophers the freedom to apply the tools of their trade—such as logical analysis, disambiguation, and drawing distinctions—to the conceptual issues that they find in science.

The third factor behind the rise of philosophy of biology stemmed from a broader trend in Anglophone philosophy. This was the move to 'naturalize' philosophical enquiry by trying to integrate it with empirical science. Traditionally philosophers have used an 'armchair' method to address the questions that interest them, such as the nature of morality, the limits of human knowledge, and the problem of free-will. A typical application of this method involves examining what a concept means and studying its logical relations to other concepts. This method has its successes, but critics have long complained that it often leads nowhere. In the mid-20th century, scientifically minded philosophers such as Willard van Orman Quine argued that empirical science could shed light on these age-old philosophical questions. It was a mistake, Quine argued, to treat philosophical questions as fundamentally different in kind from scientific ones, and thus a mistake to use exclusively armchair methods to tackle them. Opinions differ about the merits of this 'naturalistic turn' in philosophy, but it did help usher in a generation of philosophers able and willing to draw on empirical science as a resource for tackling philosophical problems.

Biology, particularly evolutionary biology, played a key role in this development. Darwin himself had predicted that his theory of evolution might have ramifications for philosophy. ('He who understands baboon would do more for metaphysics than Locke', Darwin wrote, in an allusion to the 17th-century English philosopher John Locke.) The naturalistic turn saw Darwin's prediction partly fulfilled, as evolutionary ideas were brought to bear on diverse philosophical issues. One example is the problem of intentionality. As philosophers use the term, intentionality is an attribute of mental states such as beliefs or judgements, namely that of being 'directed towards', or *about*, items in the external world. Thus if I believe that Brazil is the largest country in South America, then my belief is *about* the country Brazil. This is sometimes expressed by saying that my belief has 'representational content'—it represents the world as being one way rather than another. The hallmark of having representational content is the possibility of *mis*representation. I may represent the world as being one way when in fact it is another, that is, my belief may be false. Philosophers have long regarded intentionality as a puzzling phenomenon, for it is hard to see how it can arise in a purely physical world. After all, mental states presumably depend ultimately on the brain, and thus on neurons and their interconnections, but neurons do not seem to be 'about' anything, nor to have representational content. So how can intentionality fit into the world that modern science describes?

In the 1980s, the philosopher Ruth Millikan suggested an ingenious solution to this puzzle by drawing on Darwinism. To illustrate her basic idea, consider the honey bee's waggle-dance. This is the complex figure-of-eight dance that honey bees use to signal to their hive mates the location of a food source. Since the bee's dance has been shaped by natural selection for a particular purpose—correctly indicating where the food is—this allows us to discern a kind of proto-intentionality in the waggle-dance. We can sensibly say that a particular dance routine *means that* the food is located 30 metres away in the direction of the sun, in the sense that the biological function of the dance is to induce its hive mates to fly to this location. The bee's waggle-dance is thus capable of misrepresentation—for the food may not actually be in this location, for example if the bee has accidentally performed the wrong routine. In short, Millikan's idea is that representational content may be rendered scientifically respectable by reducing it to biological function, a notion which plays a *bona fide* role in evolutionary biology. This bold attempt to naturalize intentionality is controversial, but it illustrates how a biological perspective can help illuminate an old philosophical issue.

To summarize, the philosophy of biology emerged as a distinct field of enquiry from the interplay of three factors: the need for a less physics-centric picture of science; the presence of conceptual issues within biology itself; and the naturalistic turn in general philosophy. This book introduces philosophy of biology in a way that presumes no specialist knowledge, philosophical or scientific. The focus is mainly on evolutionary biology and genetics, as these are the areas of biology that have traditionally attracted the most philosophical interest. In recent years this situation has changed somewhat, as philosophers of biology have turned their attention to areas such as developmental biology, immunology, and microbiology. These exciting developments have opened up new avenues for philosophical reflection on the biosciences.

The structure of the book is as follows. Chapter 2 outlines the theory of evolution by natural selection, explaining its unique status in biology and its philosophical significance. Chapter 3 explores the concept of biological function and examines the controversy in evolutionary biology over 'adaptationism'. Chapter 4 examines the levels-of-selection question, which asks whether natural selection acts on individuals, genes, or groups. Chapter 5 discusses biological classification, focusing on whether there is a 'right' way to assign organisms to species, and species to higher taxa. Chapter 6 examines philosophical issues in genetics, with a focus on the concept of the gene itself. Chapter 7 examines the implications of biology for humans, asking whether human behaviour and culture can be explained in biological terms.

Chapter 2
Evolution and natural selection

In 1859 Charles Darwin published *On the Origin of Species*, in which he set out his theory of evolution. The book marked a turning point in our understanding of the natural world and was to revolutionize biology in the decades that followed. Darwin's central claims were three. First, species are not fixed but rather change their characteristics over time as they adapt to environmental conditions. Second, current species have descended from one or a few common ancestors, rather than being separately created by God. Thus an evolutionary process must have occurred, in which ancestral life forms were somehow transformed into modern ones. Third, *natural selection* is the main means of evolutionary modification. By natural selection Darwin meant the preservation of those organisms in a population who are best able to survive and reproduce in the environment, and the elimination of others. The cumulative effect of natural selection over many generations would gradually adapt organisms to their environment, Darwin argued, and eventually give rise to entirely new life forms.

Darwin was not the first to suggest that species change their characteristics over time, nor that current species have common ancestors. These ideas had been mooted by earlier scientists, including Darwin's own grandfather Erasmus Darwin. However, they were not widely accepted, because no plausible *mechanism* had been proposed to explain how evolutionary change might come about. Darwin's major innovation was to describe such a mechanism, namely natural selection, and to argue convincingly that it can lead organisms to evolve novel characteristics, ultimately leading to the evolution of new species. *The Origin* makes a persuasive case for the power of natural selection to produce evolutionary modification. Darwin offers an array of facts and observations which cannot easily be accounted for by the hypothesis that species have been separately created, but make perfect sense if his theory is true.

Natural selection is a simple but profound idea. At root, it boils down to the following logical point. Consider any population of organisms. So long as three conditions are satisfied, the population will evolve over time, in the sense that its composition will change. First, the organisms must *vary* with respect to some of their phenotypic traits—they cannot all be identical. (A 'phenotypic trait', or simply a 'trait', refers to any observable attribute of an organism, for example its height, skin colour, or skull shape.) Second, the trait variation must lead to variation in 'fitness', or the ability to survive and reproduce. This arises because some variants are usually better suited to the environment than others. For example, in a dense forest, taller plants will get more sunlight than shorter ones. Third, offspring must tend to resemble their parents. Taken together, these conditions imply that in future generations, the traits associated with higher fitness will become more common in the population, while those associated with lower fitness will decline. Darwin devotes considerable space in *The Origin* to arguing that actual biological populations do typically satisfy the three conditions.

Darwin arrived at the idea of natural selection by reading the work of Thomas Malthus, a Victorian demographer. Malthus had argued that human population growth would always outstrip the food supply. Similarly, Darwin argued that all organisms face a 'struggle for existence', since in any generation more organisms are born than can possibly survive. This is because of resource constraints—a biological population cannot grow indefinitely. So there will inevitably be winners and losers. Now from his field observations, Darwin knew that the organisms in a population typically vary in innumerable respects. It is therefore likely, he reasoned, that some organisms will have traits that confer an advantage, however slight, in the struggle for life. These organisms will survive and reproduce, transmitting the favourable traits to their offspring.

To illustrate the power of natural selection, Darwin drew a parallel with the artificial selection practised by animal breeders. This refers to how breeders modify their stock by continually selecting organisms with desirable characteristics to found the next generation. It can work surprisingly quickly, as shown by the numerous breeds of domestic dog that look quite unalike after a few hundred generations of selective breeding. In artificial selection there is a conscious agent—the human breeder—who deliberately chooses some variants over others for their own purposes. But Darwin argued that something similar goes on in nature, over a longer timescale. In natural selection there is no conscious agent, of course; rather the selective filter is imposed by the environment, favouring those organisms best able to survive and reproduce in it. This results in organisms that are well-adapted to their environment, and it generates biological diversity as different populations adapt to different environmental conditions.

The argument from design

One immediate consequence of Darwin's theory was to undermine the 'argument from design', a traditional philosophical argument for the existence of God. The most famous version of the design argument was due to William Paley, an 18th-century clergyman. Paley noted that organisms exhibit a striking 'fit' to their environment, and are functionally complex. To illustrate fit, consider a desert cactus's ability to store water, a stick insect's resemblance to the background foliage, or a bird's aerodynamic wing profile. In each case, the organism has traits that seem perfectly suited to its habitat. It is as if an engineer with foreknowledge of the environmental challenges the organism faces had deliberately equipped it with suitable traits. To illustrate complexity, consider an organ such as the vertebrate eye. As Paley noted, the eye's operation depends on the coordinated activity of many sub-parts—lens, cornea, retina, and so on—that are very finely adjusted. The eye is not unique in this respect; even a comparatively simple organism such as an amoeba contains numerous intra-cellular components that need to work together, in a coordinated way, so that vital functions like respiration and metabolism are performed. For Paley, both fit and complexity are evidence of design by a conscious deity.

Paley developed his argument with a memorable analogy. He imagined walking on a heath and stumbling across two items: a stone and a watch. Our response to the two finds would be different, he argues. The stone's presence is unremarkable—it could conceivably have always lain on the heath. But we would readily infer that the watch must have had a maker—a conscious agent who designed it. The reason is that unlike the stone, the watch exhibits functional complexity. Its internal parts are carefully adjusted to produce regular motion, and this motion serves a clear purpose: tracking the time of day. We would never seriously entertain the hypothesis that purely physical forces, such as the wind, had by chance led all the internal parts of the watch to come together in exactly the right conformation; this seems fantastically improbable. Now just as the watch must have had a maker, Paley argues, so we can infer that living organisms must have had a maker too, for they exhibit a level of functional complexity comparable with that of any human-made artefact.

Though Paley's argument had been criticized before, notably by the Scottish philosopher David Hume, it was only with the advent of Darwin's theory that it could be fully overcome. For what

Darwin provided was a naturalistic explanation of the phenomena that Paley took as evidence of the creator's handiwork. ('Naturalistic' means without appeal to supernatural or theistic causes.) Both the fit of organisms to the environment and their complexity, Darwin argued, arise from natural selection. That is, the continual preservation of the best variants in a population, and the elimination of others, creates the *appearance* of design in nature. But in reality there is no designer; rather there is just a brute causal process—natural selection—that gradually modifies organisms over time.

Modern evolutionary biology has amply confirmed the Darwinian explanation. In countless cases, biologists have pieced together a detailed picture of how organisms evolved their complex internal organization and their fit to the environment. Indeed, Paley's example of the eye—still sometimes cited as a problem case by opponents of evolution—is a case in point. Thanks to a combination of genetic analysis and studies of eye development across species, we now understand the likely sequence of stages by which the modern vertebrate eye evolved from a primitive light-sensitive organ. We know that by 500 million years ago, an essentially modern eye had evolved in the common ancestor of all the vertebrates; it was then tweaked and refined by natural selection in different species to suit environmental demands.

Though Darwin dealt a fatal blow to the design argument, in one respect he actually agreed with Paley. Darwin agreed that the organismic features that Paley pointed to are real and require special explanation. (Most modern biologists also agree with this.) There is indeed a fundamental difference between an inanimate object such as a stone and a living organism, and it is perfectly legitimate to insist on some explanation of the difference. In other words, Paley and Darwin agreed that *apparent* design is a real phenomenon, not just a figment of our imagination nor an illusion that we project onto nature. But they disagreed about its explanation. Where Paley took apparent design as indicative of *actual* design, Darwin attributed it to natural selection.

In recent years a souped-up version of Paley's argument has been resuscitated by advocates of the 'intelligent design' movement in the US. They argue, like Paley, that living organisms have features that can only be accounted for by the hypothesis that a conscious agent created them. Proponents of intelligent design often cloak their ideas in a scientific veneer, arguing that modern biochemistry reveals that cells—the basic building blocks of all organisms—exhibit 'irreducible complexity' so could not have arisen by natural selection. However, no serious biologist accepts this argument, because it ignores the extensive empirical evidence that cells did in fact evolve; and the underlying motivation behind the intelligent design movement is clearly religious rather than scientific. While Paley's argument was perfectly plausible in its day, given that he was writing without the benefit of the Darwinian theory, the proponents of intelligent design have no such excuse.

Neo-Darwinism

Darwin's theory was remarkable but not complete, for it relied on two key assumptions whose justification only became clear later. First, Darwin's theory requires that there is an ongoing supply of variation. For natural selection to operate, it is essential that the organisms in a population vary. However, selection itself is a homogenizing force—it continually *reduces* variation by preserving the best variants and eliminating others. So if natural selection is to work over a long timescale, a continual injection of new variants into the population is needed. But where does it come from? Second, Darwin assumed that offspring will tend to resemble their parents; he called this the 'strong principle of inheritance'. Such resemblance is essential if natural selection is to modify a population in the manner Darwin described. If the taller plants in a population have a survival advantage over shorter ones, this will only have a lasting evolutionary effect if tall plants tend to give rise to tall offspring. Otherwise the effect of selection will be transitory, lasting a mere generation. But what accounts for this parent-offspring resemblance?

Today we know the answer to both questions thanks to discoveries in genetics, a science of which Darwin knew nothing. Most organisms develop from a single cell, which in sexually reproducing species is formed by the union of two gametes, one each from the mother and father, into a zygote. (The male and female gametes are the sperm and egg cells.) These gametes contain genes that are transmitted from the parent. The genes in a zygote have a systematic effect on the traits that the organism develops. In a nutshell, that is why the parent-offspring resemblance, which Darwin took as a given, obtains. The injection of variation arises from two factors. First, sexual reproduction continually produces organisms with new combinations of genes; this is because an organism transmits only half of its genes, chosen at random, to a gamete. Second, genes are not transmitted with perfect fidelity. Sporadic mutations occur when the genetic material is copied, so a zygote will sometimes contain novel genetic variants not found in either parent. Mutation thus provides a continual supply of raw material for natural selection to act on.

Most genetic mutations either have no effect on an organism or are harmful. However, occasionally mutations arise that are beneficial, for example by making an organism more resistant to infection, or less likely to die in infancy, or more attractive to mates. Such mutations will be favoured by natural selection, as organisms carrying them will leave more offspring than those without, on average. Over many generations this will lead a species' genetic composition to change, as beneficial mutations become more common. The net effect of this is that the species will become better adapted to its environment, just as Darwin had posited.

Genetic mutation is often described as 'random', a term that is slightly misleading. It does not mean that mutations have no causes, nor that all genes in an organism are equally likely to mutate, neither of which is true. Rather, it means that mutations are *undirected*, in that whether a given mutation would be beneficial or harmful to an organism, in a particular environment, has no effect on the chance of that mutation occurring. That is, mutations simply occur when and where they do; their effect on an organism's fitness dictates whether they spread by natural selection, but not whether they occur in the first place.

The integration of Darwin's theory with genetics, in the early 20th century, gave rise to neo-Darwinism, which holds that random mutation and natural selection are the twin drivers of evolutionary change. Indeed, neo-Darwinists often *define* evolution as a change in a population's genetic composition, on the grounds that all other evolutionary phenomena, such as the production of new species, stem ultimately from such changes. Another neo-Darwinist tenet is the rejection of *Lamarckian inheritance*. This refers to the idea, associated with the 18th-century evolutionist Jean-Baptiste Lamarck but also believed by Darwin, that acquired characters can be inherited. Acquired characters are ones that an organism gains during its lifetime due to external influences; they contrast with innate characters that have a genetic basis. Neo-Darwinists argued that since acquired characters do not get encoded in the genes, they will not be inherited. Empirically this is generally true—a bodybuilder who pumps iron all day will not produce more muscular children as a result. However, Lamarckian inheritance is not impossible, and a number of cases have recently been documented. Biologists today recognize that genes are not the only factors that offspring inherit from their parents; others include hormones and nutrients, symbionts such as gut bacteria, learned behaviours, physical structures such as nests, and 'epigenetic' changes to DNA that affect gene expression (the process by which genes are used to make the proteins that cells need).

Modern evolutionary biology grew out of neo-Darwinism but has gone much further, in part by incorporating the findings of molecular biology (see Chapter 6). Today, we have a deeper understanding of the mechanisms of evolution, of how new species arise, and of the molecular basis of many of the adaptive traits that natural selection has produced. Moreover, we have a detailed knowledge of what the tree-of-life looks like, that is, the pattern of descent that links all modern species to a single common ancestor. And we have some knowledge, though less than we would like, of the earliest life forms on earth and how they came into existence. The ensuing picture is more complicated than the one that Darwin painted, but his fundamental insights into the evolutionary process, in particular the role of natural selection in modifying organisms and in creating diversity, remain intact.

The logic of Darwinian explanation

Darwin's explanation of how organisms became adapted to the environment is interesting philosophically, because it involves a distinctive logic. To appreciate this logic, it is useful to contrast Darwin's explanation with Lamarck's. Lamarck argued that individual organisms can adapt to the environment *within their own lifetime*. Thus for example, Lamarck argued that giraffes' necks became elongated because ancestral giraffes stretched to reach tall trees, which caused their necks to lengthen; they then transmitted their modified necks to their offspring. Now since acquired characters are not usually transmitted, as we know, Lamarck's explanation is empirically suspect. But that is not the point here. Rather, it is the logical contrast between Lamarck's and Darwin's explanations that is of interest. Lamarck's involves *individuals* changing to improve their adaptive fit to the environment, whereas Darwin's treats the *population*, rather than the individual, as the unit that changes. No individuals undergo adaptive change on the Darwinian theory; rather the population changes, via the selective preservation of some variants and the elimination of others.

The biologist Richard Lewontin expressed this by saying that Lamarck's theory is *transformational* while Darwin's is *variational*. This distinction applies more widely than in biology, as an example due to Elliott Sober illustrates. Suppose we find that the schoolchildren in a particular class are unusually good at music and wish to know why. One possible answer would focus on each individual child and seek to explain their musical ability. For example, we might find that each child grew up in a musical family, started piano lessons early, had parental encouragement, and so on. This is a transformational explanation—it explains how each child was transformed into an able musician. But there is a second quite different way to answer the question. We could point out that in order to be admitted to the class in the first place, children had to score top marks on a difficult music test. That is, the class's membership was determined by a selection process which filtered children according to their musical ability; as a result, the class contains only children who are good at music. This is a variational explanation.

In the schoolchildren example, the two explanations explain subtly different things, as Sober observes. The first explains why the children in the class have high musical ability, rather than those *self-same* children having lesser ability. The second explains why the class as a whole contains musically able children, rather than containing *other* children who are less able. So the former explains facts about individual children, while the latter explains a fact about the composition of the class. In just the same way, Darwinian explanations in biology explain population-level, not individual-level, facts. For example, Darwinism does not explain why any individual polar bear has a white coat as opposed to a brown coat. To explain that, we would point to the fact that the bear in question was born with genes that caused it to develop a white coat—which is a developmental, not an evolutionary, explanation. What Darwinism does explain is why the polar bear population (or species) contains white-coated individuals, as opposed to containing *other* brown-coated individuals; or equivalently, why the frequency of the trait 'white coat' in the polar bear species is 100 per cent.

Ernst Mayr, a leading 20th-century evolutionist, argued that Darwin had invented a new way of thinking about nature which he called 'population thinking'. This is not a precisely defined term; but in part, it means treating the population as the relevant unit of analysis and regarding variation within the population as a 'real' feature of it, rather than mere noise. Mayr is right that this is an integral aspect of Darwinism. As we have seen, the existence of variation is a prerequisite for natural selection to operate, and Darwinian explanations are concerned with change in a population, not in a single individual. Moreover, modern evolutionary biologists devote much effort to studying and quantifying population-level attributes, such as the extent of genetic variation within a population, and the magnitude of the fitness differences between variants. Thus part of Darwin's legacy was indeed to instigate a conceptual shift in which populations, as well as individuals, became objects of study in their own right.

Proximate and ultimate questions

The theory of evolution occupies a unique place in modern biology. It functions as a grand organizing principle that brings order to a plethora of empirical facts. Modern organisms bear indelible traces of their evolutionary past, including genetic signatures, adaptive traits, and cross-species similarities that reflect common ancestry. The biologist Theodosius Dobzhansky wrote in 1973 that 'nothing in biology makes sense except in the light of evolution', an assertion that could equally be made today. Of course, much biological work is not explicitly evolutionary. An ecologist studying the biodiversity in a rainforest, a microbiologist trying to clone a virus, or a primatologist studying chimpanzees in the wild are concerned with events in the here and now, not the distant past. One might think that such day-to-day research could proceed identically even if Darwin were wrong and the story of biblical creation were right. To an extent this is true, but in fact evolution forms the backdrop to many biological enquiries, even if their overt focus is with the present.

Why is evolution so important for biology? One answer is that it provides a unique type of understanding that would otherwise be unattainable. To see this, let us focus on a famous distinction, again due to Mayr, between two different sorts of question that biologists ask. *Proximate* questions ask how a particular biological mechanism works. For example, how does a migrating salmon find its way back to its natal river (or 'home')? How does a mammal regulate its body temperature? How does a bacterium move towards ambient oxygen? Such questions are answered by describing the causal factors that give rise to the phenomenon in question. In the salmon example, the answer is that juvenile salmon imprint on the odour of their natal stream, then draw on this memory to navigate back there as adults. In the bacterium example, the answer is that bacteria have receptors that sense the concentration of chemicals in the environment; by rotating their flagellar motor they can swim through a chemical gradient.

Ultimate questions, by contrast, are concerned with evolutionary advantage rather than proximal mechanism. They typically ask 'why' rather than 'how'. For example, why do salmon migrate back to their natal river, rather than staying put? This is a good question to ask, since homing is energetically costly, time-consuming, and potentially dangerous. Why then do salmon do it? This question cannot be answered by studying the mechanistic details of how salmon navigation works, interesting though they are. Rather, it requires us to find the evolutionary advantage of the salmon's homing behaviour, that is, the reason why natural selection led it to evolve in the first place. For example, one plausible hypothesis is that by returning to its natal stream, a salmon can locate a habitat favourable for spawning and juvenile survival, thus gaining a fitness benefit. If this hypothesis is correct, as many biologists believe, then it tells us *why* salmon do what they do, rather than *how* they do it.

Proximate questions are largely independent of evolutionary history. Even if Darwin were wrong and salmon had been created by God, the question of how they navigate back to their natal stream could still be posed, and answered, in essentially the same way. But ultimate questions are not like this. Rather, a typical ultimate question presupposes that the trait in question is an adaptation, that is, has evolved by natural selection because it confers a fitness advantage. To answer the question of why the organism has the trait, the biologist will seek to identify what this advantage is, that is, to give an adaptive explanation. This involves making a claim (at least implicitly) about the course of evolutionary history. A biologist who explains why salmon home by saying that it is to locate a favourable spawning ground is making a claim about why natural selection led the homing behaviour to evolve. The explanation would have to be rejected if the theory of evolution turned out to be untrue.

The centrality of evolution to biology arises because, by virtue of explaining how adaptations arise, evolutionary theory is uniquely able to answer the ultimate 'why' questions. Fields such as molecular biology, cell biology, and developmental biology supply detailed knowledge of how organisms and their sub-components work. Such knowledge is scientifically invaluable and of great interest in its own right. But evolutionary biology adds a further and fundamentally different layer of scientific understanding. It allows us to make sense of the variety of organismic attributes that we observe, by explaining them as 'rational' or adaptive responses to the environmental challenges that organisms face. This is complementary to, but quite different from, the type of understanding that we get from the study of proximate mechanism, and is the source of much philosophical fascination with evolutionary biology.

Though proximate and ultimate questions must be sharply distinguished, in some cases the answers to them are intertwined. Knowledge of the evolutionary past may contribute to our understanding of how organisms work today. The field of Darwinian medicine illustrates this point. Medicine's main concern, of course, is with how the human body works today and with the proximate causes of disease. But proponents of Darwinian medicine argue that an evolutionary perspective can aid our understanding of this. Consider for example obesity. Evolution teaches us that humans' food preferences evolved in a very different environment to the one we live in today, in which nutrients were in short supply and a craving for sugary food was beneficial. So there is a mismatch between our current environment and our evolved food preferences, which explains why humans are prone to overeat. If this explanation is correct, as is widely believed, it may have practical implications for how to counter the obesity epidemic, or for assessing the likely success of a particular intervention. So evolutionary considerations may help us to better understand proximate mechanisms.

Why believe in evolution?

The theory of evolution is a mainstay of modern biology, and no biologist today seriously disputes its truth. Despite this, in society at large one often finds a marked reluctance to believe in evolution, even among educated people who are happy to accept the rest of the modern scientific worldview. (This reluctance is stronger in some countries than others.) Why is this? Three different factors seem to be at work.

First, evolution is a distinctly unsettling idea that can be hard to accept at first encounter. If you look at a slug on a strawberry plant in your garden, for example, it takes a considerable leap of imagination to accept that you, the slug, and the plant all share a common ancestor if we trace back far enough. And yet evolutionary theory tells us that this is true—the ancestor of all plant and animal life was a single-celled protist (similar to algae) estimated to have lived some 1.6 billion years ago. So there is a disconnect between our everyday experience of the living world, which presents us with an array of life forms that seem categorically distinct, and the central claim of evolutionary biology, which is that they are all related by common ancestry.

Second, evolution dethrones humans from the unique position that tradition had accorded us. For centuries it was believed that humans were fundamentally different from other living creatures, or even outside the natural order altogether. For example, Descartes argued that non-human animals were mere automata, while humans have souls. The belief in human exceptionalism, in some form, is still widespread. But evolution undermines this belief. It tells us that *Homo sapiens* is just another species of primate that diverged from the chimpanzees a mere six to eight million years ago. Of course, humans do have distinctive attributes, such as language and culture, that most other species do not, but this is against a background of commonality. From an evolutionary perspective, the idea of a fundamental discontinuity between humans and the rest of the living world is an illusion.

Third, and most obviously, evolution conflicts with much religious doctrine, particularly of the Abrahamic religions. For example, the Book of Genesis says that living creatures were created separately by God a few thousand years ago, over a six-day period. Taken literally, this claim obviously conflicts with what evolution teaches us, namely that life on earth evolved gradually over a period of about four billion years. So it is unsurprising that from Darwin's day to the present, much of the opposition to evolution has had a religious motivation, whether explicit or covert.

These three factors explain the resistance to evolution, but do not of course justify it. The fact that a scientific idea conflicts with our everyday experience, or punctures human vanity, or contradicts religious doctrine is not a good reason to reject it. But it does prompt the question: how can we be sure that the theory of evolution is actually true? How strong is the evidence in its favour?

To address this matter, we need to be more precise about what we mean by the theory of evolution. We have seen that Darwin himself argued for three propositions: (i) species evolve new characteristics over time; (ii) current species have descended from common ancestors; and (iii) natural selection is the main means of evolutionary modification. The evidence for each of (i)–(iii) is somewhat different. Let us focus here on (ii), the claim of common ancestry, as this is what opponents of evolution are primarily concerned to deny.

The modern form of proposition (ii) is the one-tree-of-life hypothesis, which says that all current species can be traced back to a single common ancestor. The main reason for believing this is essentially the same as in Darwin's day: it can explain a vast number of facts that would otherwise be inexplicable, or would have to be accepted as brute coincidences. Darwin himself cited, among many other facts, the striking anatomical similarities between different species, such as the limbs of horses and cows. Such similarities make immediate sense in the light of evolution. Horses and cows have a common ancestor from which they inherited their quadrupedal body plan, which then underwent subsequent evolutionary modification. But if instead horses and cows were created separately, so had no common ancestor, we would not expect their anatomy to be similar at all.

Another set of facts that bear witness to common ancestry, also noted by Darwin, comes from embryology. Consider the fact that the embryos of all vertebrates are so similar, in their early stages, that they cannot easily be told apart. As the embryo develops, traits appear that are characteristic of its vertebrate sub-group, such as birds, and eventually of its species. Again, this makes good sense in the light of evolution. All vertebrates trace back to a single ancestral species that evolved into distinct lineages, each with their own characteristic features: wings for birds, gills for fish, and so on. These features develop after the basic vertebrate body plan has been laid down, for it is hard to make a change in early embryology without adverse effects on the whole organism. As a result, the embryos of all vertebrates are similar in their early stages, diverging as embryonic development proceeds. There would be no reason to expect such similarities if vertebrate species had been separately created.

The most powerful reason for believing the one-tree-of-life hypothesis was not available to Darwin. It is the *universality of the genetic code*. To understand this, we need some basic genetics. We can think of a gene as a set of instructions, inscribed in DNA, for building a single protein. A protein consists of a long chain of amino acids linked together. The genetic code refers to the way in which a gene's DNA sequence determines the sequence of amino acids in the corresponding protein. (More precisely, the code maps each triplet of nucleotides onto a single amino acid; see Chapter 6.) There are a vast number of possible genetic codes, all of which are compatible with the laws of chemistry. But here is the striking thing: all living organisms, from flies to bacteria to humans, share the *same* genetic code (or very nearly). This makes perfect sense if all organisms have a common ancestor but would be an astronomical coincidence otherwise. It would be akin to the very same human language arising independently on numerous separate occasions. If anthropologists discovered a number of isolated tribes who speak exactly the same language, they would obviously assume that the tribes are offshoots of a single linguistic community. The alternative supposition that each tribe independently invented the same language is phenomenally implausible, given the indefinitely large number of possible languages. Similarly, the universality of the genetic code among today's organisms provides overwhelming evidence of their common ancestry.

It is sometimes said, particularly by those who dislike evolutionary biology, that evolution is 'just a theory'. Some care is needed in interpreting this statement. Certainly there is such a thing as 'the theory of evolution', that is, a set of theoretical principles that explain how evolution works, some of which we have outlined above. But the statement is usually taken to mean that evolution is not an established fact but rather a speculative hypothesis which a rational person might dispute. This idea should be resisted. Clearly we cannot directly observe the past, so the claim that current species have descended from common ancestors is something that we can only know indirectly, by inference. But the same is true of many scientific propositions. We cannot directly observe electrons, nor ancient civilizations, nor the sun's core, but we know plenty about each of these things. The evidence in favour of evolution is so overwhelming that there can be no serious doubt as to its truth. Anyone who doubts the reality of evolution on the basis that it is 'just a theory' should, to be consistent, doubt virtually every other proposition of modern science too.

Chapter 3
Function and adaptation

A striking feature of the biological sciences is their frequent appeal to the notion of *function*. Consider the following statements. The function of the crab's shell is to protect its innards; the function of the human kidneys is to cleanse the blood; the function of the male bird-of-paradise's dance is to attract mates. Statements of this form, which we may call 'function-attributing statements', can be found in almost any biology textbook, and often pass without notice. This marks an interesting contrast with the non-biological sciences, where talk of function is virtually non-existent. Geologists do not talk about the function of glaciers; astronomers do not talk about the function of planets; and chemists do not talk about the function of covalent bonds. Indeed if they did talk this way, it would be hard to know what they meant. Why then do biologists make such liberal use of function-talk, and what exactly does it mean?

A naïve answer to this question is that to speak of the function of some biological item is simply to say what it does, or to describe its effects. On this view, to say that the function of the crab's shell is to protect its innards means just that the shell does protect the innards; to say that the function of the bird's dance is to attract mates means just that the dance does attract mates; and so on. However this cannot be quite right. For after all, the crab's shell has other effects as well as protecting the innards, for example making it easy for fishermen to pick crabs up. But we would not describe this as the shell's function. Similarly, when a male bird-of-paradise performs its courtship dance it will often attract predators as well as mates; but attracting predators is not the function of the dance, it is just something that the dance happens to do. As these examples show, not all of a biological item's effects are part of its function; to think otherwise is to miss the distinction between genuine function and what we might call 'unintended side-effect'. Biologists implicitly invoke this distinction whenever they use the functional idiom.

At this juncture a philosophical puzzle arises. For it can easily seem that function-talk in biology has a *normative* dimension, resting on value judgements. (In philosophy, the normative contrasts with the descriptive; the former deals with how things should be, the latter with how things actually are.) When a biologist talks about the function of some biological item, they appear to be saying something about what the item is *meant* to do, rather than what it does do. Indicative of this is that wherever talk of function makes sense, so does talk of malfunction. A shell that does not protect the crab's innards, or a kidney that does not cleanse blood from the body, has failed to do its job, or malfunctioned. But this is rather puzzling. Who gets to decide what some biological item is meant to do? The natural sciences, surely, are in the business of describing the world, not making judgements about how it should be. Is biology really an exception to this principle? Or is there a way to explain away the normative dimension, perhaps by translating function-attributing statements into other terms? We return to this puzzle below.

You might wonder whether all of this is just fussing about a word. The answer is no, for two reasons. First, it is not the single word 'function', but rather a broader family of idioms used in biology that raise the same philosophical issue. Other members of the family include 'for' and 'in order to'. Thus the plant's stamen is *for* making pollen, and the swallow migrates south *in order to* escape the harsh winter. Second, the interesting phenomenon here is not just about language, but also about a characteristic pattern of explanation. The pattern involves explaining the existence, or the nature, of some biological item by saying what the item is for, that is, what its function is. For example, suppose we ask why cacti have spines. Answer: the cactus's spines are for deterring herbivores. Or suppose we ask why fish have gills. Answer: fish have gills in order to breathe underwater. As these examples show, citing the function of a biological item often plays a key role in the explanations that biologists give.

Such functional explanations have a clear affinity to the ultimate (or adaptive) explanations examined in the last chapter, in which an organism's traits are explained by showing how they contribute to its biological fitness. Indeed, one might think that the two sorts of explanation—functional and adaptive—are in fact one. On this view, the function of some biological item is simply its adaptive significance, that is, its specific contribution to the organism's fitness; and to explain the item's existence, or its nature, by citing its function is equivalent to giving an adaptive explanation, that is, to saying why natural selection led to its evolution. This is an attractive view, but as we shall see, it is not the only proposal for how to understand function-talk in biology.

Functions and artefacts

We observed that the physical sciences have little use for the notion of function. However, there is one non-biological context where function-talk is common, namely in the description of human-made artefacts. For example, we say that the function of the sundial is to tell the time; the function of the keel is to stabilize the boat; and the function of the valve is to maintain the tyre's air pressure. Now of course artefacts, unlike biological organisms, have been intentionally designed by humans to perform a particular task. And this fact seems crucial to understanding what function-talk means in this context. What makes it true that a sundial is for telling the time is that someone deliberately designed the sundial with the intent that it should keep time accurately. Similarly with the boat's keel and the tyre's valve. Thus the function of an artefact, or an artefact part, derives from the intention of its human designer.

Artefact functions share interesting similarities with biological functions. First, the distinction between function and unintended side-effect applies equally to artefacts. A boat's keel does many things as well as stabilizing the boat, such as providing a home for barnacles and a place from which to hang the rudder. But these are not the keel's function—they are not what boat-builders intended the keel to be used for. Second, the concept of malfunction also applies to artefacts. A sundial that does not track time accurately, or a valve that leaks, is defective: it is not working as it should. Thus a notion of 'correct' functioning applies to artefacts, where this means 'in accordance with the intentions of the artefact's maker'. Third, as in the biological case, citing an artefact's function can serve to explain why it exists, or why it has the features it does. Suppose a child asks why boats have keels and why pneumatic tyres have valves. A reasonable answer would be that keels are for stabilizing the boat, and valves are for maintaining tyre pressure.

Given these similarities, one might wonder whether talk of biological function is simply a hangover from pre-Darwinian times, when it was believed that organisms, like artefacts, had also been intentionally designed. On a creationist worldview, attributions of function could naturally be understood as referring to the creator's intentions (presuming they could be discerned). The statement that the function of the crab's skeleton is to protect its innards could be interpreted to mean that God had designed the skeleton for the purpose of innards-protection; and similarly for other function-attributing statements. So perhaps function-talk in contemporary biology is merely a casual inheritance from a previous era, similar to how astronomers talk about 'sunrise' despite no longer believing that the sun goes round the earth.

This is a coherent suggestion, and it would certainly explain the similarities between artefact functions and biological functions. But it is not especially plausible. For when contemporary biologists employ the functional idiom they give every impression of talking seriously, and saying something that has a precise, non-metaphorical meaning. Unless this is so, it is hard to make sense of the many debates in biology about the functions of particular traits. For example, the human appendix was traditionally said to have no function; but recent research has called this into question, arguing that the appendix does in fact have a function, namely to serve as a reservoir for beneficial gut bacteria. The biologists involved in this research employ function-talk quite deliberately; it is hard to believe that their language is merely a vestige of a pre-evolutionary worldview.

An alternative suggestion is that the similarities between artefact and biological functions arise because natural selection plays a role akin to that of a human designer. This has some plausibility, since as we saw in Chapter 2, evolved organisms often exhibit *apparent* design, given their remarkable adaptive complexity. The suggestion, then, is that we can sensibly talk of an item's function when one of two conditions is met. Either the item has been intentionally designed by an agent in order to perform a task, or else the item exhibits apparent design, that is, has evolved by natural selection because of its effect on fitness. If correct, this would explain why it makes sense to talk about the function of a kidney or a sundial, but not of a glacier. This suggestion leads us straight to the most popular philosophical analysis of function-talk in biology: the aetiological theory of function.

The aetiological theory of function

The 'aetiology' of something means its causal history. The basic idea behind the aetiological theory is simple, namely that to talk of an item's biological function is equivalent to saying something about why natural selection led that item to evolve. Consider again the bird-of-paradise's dance. It is a plausible hypothesis that selection led the bird's dance to evolve, and to have the precise features that it does, because it attracts mates, not because it attracts predators. This is plausible because attracting mates contributes positively to an organism's fitness while attracting predators does not. If this hypothesis is true, then the aetiological theory says that the function of the dance is to attract mates. More generally, the theory holds that the function of a biological item, or trait, is to be identified with the effect of the item in virtue of which natural selection favoured it. This theory is also known as the 'selected effect' theory of function.

For another example, consider the polar bear's white coat. This trait has a number of effects, including reflecting sunlight, camouflaging the bear when hunting, and giving polar bear cubs a cuddly appearance that human zoo goers like. Given what we know about the environment in which polar bears evolved, it is very likely that the second of these effects—camouflage—is the reason why natural selection led the polar bear to evolve a white coat. For a bear that is well-camouflaged has an obvious selective advantage over one that is not, so on average will leave more offspring. If this is correct, then the aetiological theory implies that camouflage is the function of the polar bear's coat, while reflecting sunlight and appearing cute to humans are not. This is intuitively the right result, and corresponds well with actual biological usage.

The aetiological theory makes good sense of much function-talk in biology, and it has other advantages too. It offers a principled basis for the distinction between function and side-effect, in terms of the reason why natural selection favoured the trait. It accounts for the explanatory role of function attributions, in particular the fact that citing a trait's function serves to explain why it exists. Also, the aetiological theory promises to explain away the apparently normative dimension of function-talk. For it implies that the function/malfunction distinction does not really involve value judgements about how organisms 'should' be. To say that a kidney is malfunctioning, on the aetiological theory, means simply that the kidney is not having the effect that, historically, is the reason why kidneys evolved. That is, there is an objective fact about what kidneys and other biological traits are 'meant' to do, that derives from their evolutionary history.

One challenge to the aetiological theory concerns the uniqueness of the functional attributions that it licenses. Biologists often refer to 'the' function of a trait, implying that it has exactly one. But if function is identified with selected effect, there will often be more than one candidate for the function of a trait. For example, instead of saying that the function of the polar bear's white coat is camouflage, we could equally say that its function is to allow the bear to hunt in the daylight, or to enhance the bear's hunting success, or to help prevent starvation—or simply to increase its fitness. The bear's coat does each of these things, and it seems arbitrary to identify one of them as 'the' reason why the coat was favoured by natural selection. Though this is true, the problem here is not specific to the aetiological theory of function, nor even to biology. Whenever an effect lies at the end of a long causal chain, there is a multiplicity of candidates for 'the' cause. Suppose someone asks what the cause of World War I was. Possible answers include: the imperialist ambitious of the European powers; the ensuing arms race; the shot fired at Archduke Ferdinand; and others. So how do we identify 'the' cause? This is a quite general philosophical problem, over which much ink has been spilled. (One solution says that it is a pragmatic matter, not a matter of objective fact, what we designate as 'the' cause.)

Whatever the solution to this problem, it seems likely that it will be applicable to the aetiological theory.

A different challenge to the aetiological theory comes from traits that originally evolved for one purpose but were later co-opted for another. Such traits, known as 'exaptations', arise not infrequently in evolution. For example, birds' feathers originally evolved for thermoregulation, but were later modified by natural selection to improve aerodynamic performance. What then is the function of feathers according to the aetiological theory? To keep birds warm, or to help them fly efficiently, or both? There seems no clear way of answering this question. However this is not fatal to the aetiological theory, for two reasons. First, by refining the functional attribution the problem can be partly circumvented. Bird feathers do not form a homogenous class; some such as tail feathers are specialized for flying, while others such as down feathers are specialized for keeping warm. Moreover, feathers of all sorts have multiple aspects, for example shape, texture, and size; by focusing on a particular aspect, it may be possible to narrow down the candidate functions. Second, there is anyway no consensus among biologists about how to employ function-talk in cases of exaptation. That a philosophical analysis of what such talk means does not yield an unambiguous answer is therefore not a mark against it.

Despite its advantages, the aetiological theory has one implication that many people find problematic. The theory implies that whenever a biologist makes a function-attributing statement, they are saying something, implicitly at least, about the course of evolutionary history. Now in some contexts this is plausible but in others it is harder to swallow. Consider, for example, the human heart. As every schoolchild knows, it was William Harvey, a 17th-century physician, who first said that the function of the heart is to pump blood around the body; and modern biologists agree that Harvey was right about this. But the theory of evolution had not been discovered in the 17th century, so when Harvey spoke of the heart's function, it seems clear that he did not mean 'effect in virtue of which hearts have been selected'. If we insist that the aetiological theory is correct, we seem forced to conclude either that Harvey did not understand what his own words meant, or that he was using the word 'function' in a quite different sense from modern biologists. Neither option is very palatable.

This example is a dramatic way of making a more general point, which is that some functional attributions in biology do not seem to have much to do with evolution; rather, they derive from studying how organisms work in the present. This is particularly true in the biomedical sciences. For example, researchers studying the molecular basis of a disease often try to identify the function of a biochemical pathway that is implicated in causing the disease. To this end, they make experimental studies on how the pathway works and what effect it has on cells and tissues. Such research has nothing to do with evolution, on the face of it, yet often leads to confident attributions of function. This observation motivates the main rival to the aetiological theory, namely the 'causal role' theory of function.

The 'causal role' theory of function

The key idea of the causal role theory is that functional attributions are often made in the course of a particular sort of scientific investigation, namely trying to understand how a complex biological system or process works. Consider for example thermoregulation in humans. This involves a control system in which a brain region, the hypothalamus, monitors body temperature and sends impulses to initiate appropriate physiological responses, such as sweating. The system involves a large number of specialized parts and sub-systems, each of which does a particular job. To understand how the thermoregulatory system as a whole works, we need to know the contribution that each part makes. It is here that the notion of function comes in, according to the causal role theory. The function of some part is simply its contribution to the operation of the overall system, which enables the system to do what it does. Thus the function of the thermoreceptors in the hypothalamus is to detect blood temperature, while the function of the sweat glands is to secrete sweat onto the skin's surface. It is because the thermoreceptors and sweat glands do these things that the thermoregulatory system is able to keep body temperature constant.

To take another example, consider the adaptive immune system in vertebrates, which eliminates pathogens from the body. Again, this system contains a number of parts, namely T and B cells of various types, each with a specialized role; and understanding how the immune system works requires knowing what these roles are. Thus killer T cells recognize and kill virus-infected cells, while B cells bind to antigens and produce antibodies. These then are the functions of the T and B cells, according to the causal role theory, for it is because they do these things that the adaptive immune system is able to successfully eliminate pathogens. In short, an item's function is its specific contribution, or causal role, in the operation of a larger system; while a side-effect is something that the item does that does not so contribute.

We saw that the aetiological theory emphasizes that function attributions are often answers to the ultimate question 'why does it exist?' (Why does the cactus have spines? To deter herbivores.) The causal role theory, by contrast, draws attention to a proximate question, namely 'how does it work?' (How does the immune system eliminate pathogens? By means of T and B cells, which kill intruders and produce antibodies.) The latter question occupies centre-stage in many biological investigations, particularly in fields such as physiology, cell and molecular biology. Such investigations typically involve focusing on a complex structure or process, studying the causal interactions among its different parts, and trying to determine what contribution each part makes. It is here that talk of function gets its purchase, according to the causal role theory.

The chief merit of the causal role theory is that it makes sense of the point that proved problematic for the aetiological theory, namely that function attributions are often made in the course of studying how some biological system works in the present, rather than in the course of studying evolutionary history. The causal role theory can easily account for the fact that pre-Darwinian biologists knew a lot about functions, and for the fact that modern biologists attribute functions in non-evolutionary contexts. Even if a living organism had sprung into existence fully formed yesterday, this would make no difference to the truth of a statement about the function of one of its traits, when function is understood as causal role. But on the aetiological theory the statement would necessarily be false, as the organism would lack the evolutionary history necessary for its traits to possess functions.

Having said this, in practice the aetiological and causal role theories will often agree on how to assign functions, which is why the difference between them has often gone unnoticed. Consider again the mammalian heart. It is because the heart pumps blood around the body that hearts were favoured by natural selection; and it is also true that pumping blood is the heart's contribution to the operation of the cardiovascular system, in virtue of which the system works as it does today. Similarly, it is because the cactus's spines deter herbivores that natural selection led them to evolve, and herbivore-deterrence is also the causal role that the spines play today, in the cactus's defence system. So in these cases, selected effect and causal role coincide. But this need not always be the case, and even when it is, the two theories still disagree about what *makes it true* that a trait has a given function.

One objection to the causal role theory says that it is ultimately reliant on a source of design, which must come either from a designer's intentions or from natural selection. To see this objection, note that when we explained the idea of causal role above, we talked about the 'operation' of a biological system, or how it 'works'. But surely these are covert ways of talking about what the system is meant to do? When we identify the causal role of the sweat glands in the thermoregulatory system, we are implicitly assuming that the system is meant to achieve some goal, namely keeping body temperature constant. But that assumption is only justified, the objection goes, if the system has either been consciously designed for that goal or has evolved by natural selection to achieve it. In short, we can only plausibly identify an item's function with its causal role in a larger system if that system is the result of conscious design or its surrogate, natural selection. So in the end, true function attributions in biology do rely on natural selection; thus the aetiological theory is the winner.

This is a powerful objection to the causal role theory, but it is not entirely conclusive. An alternative view is that we should simply admit both notions of function, or be 'pluralists' in philosophers' jargon. On a pluralist view, selected effect and causal role are both valid notions of biological function, each tailored to a different type of enquiry. The former is of relevance in evolutionary contexts when we are trying to answer ultimate questions about why traits evolved; the latter is of relevance when we are trying to answer proximate questions about how organisms work. Given that these two sorts of enquiry are fundamentally different, pluralists argue that it is a mistake to try to assimilate the two notions of function, or to try to choose between them; rather we should accept both, and hope that context will indicate which notion is in play. Though attractive, the pluralist view raises a number of awkward questions, such as why a single word should be used for two such different things.

Though the debate over function has been primarily pursued by philosophers, it is relevant for biological practice. An illustration of this comes from the recent controversy in biology over 'junk DNA'. Recall that DNA is the macromolecule that genes are made of. However in most species, including humans, the genes make up only a small proportion of the total DNA content in a cell. For decades, the consensus was that the bulk of the DNA in the human genome does nothing, so has no function. This consensus was challenged in 2012 by researchers from the ENCODE project. Based on a large volume of experimental data, they argued that much of so-called junk DNA does in fact have a biochemical function, playing a critical role in how cells and tissues behave. However, W. Ford Doolittle, a prominent Canadian biologist, has questioned this interpretation of the data. He argues that the ENCODE researchers are implicitly using too liberal a criterion for function, namely having some biochemical effect or other, which is tantamount to conflating function with side-effect. If we identify function with selected effect, as Doolittle recommends, then the traditional 'junk' interpretation is still defensible, he argues. Whoever turns out to be right, this example shows how a scientific debate can have a hidden philosophical dimension.

Functions and the adaptationist programme

Suppose we accept the aetiological account of what function-attributing statements in biology mean. A further question is how we can tell whether such statements are true. Given that our knowledge of evolutionary history is partial, how can we ever be sure what a trait's function is, or that it has one at all? The short answer is that we cannot be sure, but we can often achieve something close to practical certainty. For often enough, the environment in which a species evolved is similar to the one it currently inhabits; so if a trait has an obvious adaptive significance in today's environment, it is often fairly easily to identify the reason why it originally evolved, particularly if we can make relevant cross-species comparisons. For example, the polar bear's white coat is clearly useful for hunting in today's Arctic; and we know that polar bears evolved quite recently from brown bears, which have a much wider geographic distribution. So there is no serious doubt that camouflage is the reason why natural selection led the polar bear to evolve its white coat. This is not an isolated example; but it must also be admitted that there are cases where functions, in the sense of selected effects, are much harder to discern.

The issue of how reliably we can determine traits' functions was part of a major controversy in 20th-century evolutionary biology. In a famous 1979 paper, the biologists Stephen Jay Gould and Richard Lewontin launched an attack on what they called the 'adaptationist programme' in biology. This programme involves studying organisms on the assumption that they are by and large well-adapted, and that their traits have identifiable functions. Adaptationist reasoning has often been successful, since thanks to natural selection many organisms are indeed well-adapted to their environment. Gould and Lewontin did not deny this, however they accused proponents of adaptationism of behaving unscientifically. They argued that adaptationists assume without proof that every trait has a function, and simply invent stories about what the alleged function is, without sufficient evidence. (They dubbed these 'just-so stories', after Rudyard Kipling's tales for children.) Adaptationists are therefore guilty of a kind of cognitive bias according to Gould and Lewontin: they are predisposed to see function wherever they look, and they neglect the possibility that some traits may not have a functional or adaptive explanation at all.

Gould and Lewontin suggested a number of reasons why a given organismic trait may lack a function, and thus why adaptationist reasoning may lead astray. Their first reason was based on a memorable architectural analogy. A 'spandrel' refers to the triangular space between a curved arch and a rectangular frame or a dome (Figure 1). In St Mark's Cathedral in Venice, the spandrels are richly adorned with paintings of the twelve apostles. But clearly it would be wrong to think that spandrels are 'for' anything; the architect did not create them in order that they could be adorned. Rather, spandrels are an inevitable by-product of building a domed roof. Similarly, Gould and Lewontin suggested that many organismic traits may be by-products, rather than having a function of their own. The human chin is a possible example. Since great apes lack chins, we know that chins evolved somewhere in the hominid lineage. But why? Though possible

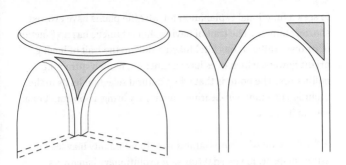

1. A spandrel is the triangular space between a curved arch and a rectangular frame or dome.

adaptive benefits of the chin have been suggested, such as helping to intimidate opponents in fights, many biologists believe that the chin is simply a by-product of how the skull and jaw bones develop in humans. If this is right then the chin is a spandrel, and thus cannot be given a functional explanation. Similarly, it has been suggested that aspects of the human mind, such as consciousness, are spandrels, arising as a by-product of increased brain size.

A second reason why a trait may lack a function is related. It stems from the fact that a single gene can influence multiple traits. As a result, there are often genetic correlations between traits, that is, an organism with trait A is likely to have trait B too. This means that if one of the traits is favoured by natural selection, the other will come along for the ride, or hitchhike. For example, suppose that in a given plant species, plants with brightly coloured flowers enjoy a selective advantage over duller ones, as they are more attractive to pollinators. Suppose also that for genetic reasons, having brightly coloured flowers correlates with having short stamens. And suppose that stamen size itself is selectively neutral, that is, has no effect on fitness. Then, if natural selection leads the flowers to evolve brighter flowers, it will also lead them to have shorter stamens. But it would be a mistake to seek a functional explanation for why the plants have evolved shorter stamens. The trait of having short stamens has no function of its own; rather it has hitchhiked a ride on the trait of having bright flowers, which does have a function, namely attracting pollinators. The point is that when natural selection leads to the evolution of a function-bearing trait, it may bring a non-functional trait in its wake.

A third reason why the assumption of functionality may lead us astray stems from vestigial traits, or evolutionary hangovers. (The human appendix is a traditional example, though as noted above the lack of functionality is unclear.) Human goose bumps are a better example. Goose bumps in humans are a degenerate form of a trait found in all mammals, in which their fur stands on end in response to either cold or fear. In many mammals this response is adaptive, since erect fur serves to trap heat and may intimidate enemies. But human hairs are too short to do either of these things effectively. Now if an adaptationist were unaware that goose bumps did not evolve *de novo* in the human lineage, they might wrongly assume that they *must* have a function in humans, and be led to advance a speculative story about what that function is. This of course would be an error. The general point here is that evolution does not design each species anew but rather works with what it has; and vestigial traits are one result of this.

This point leads to a fourth reason why adaptationist reasoning has limits, namely the existence of what are called 'developmental constraints'. The process of embryonic development, by which a zygote gives rise to an adult organism, may strongly constrain the possible variants from which natural selection chooses. Consider for example the tetrapod (four-limbed) body plan found in all land vertebrates. Since this body plan is so widespread, an adaptationist might be tempted to posit a functional reason for this, that is, to argue that four limbs is adaptively superior to any other limb number. This could be true; but an alternative explanation is simply that once the tetrapod body plan had first evolved, there was no way that evolution could undo it, given how development works. That is, there is no easy way that genetic mutations can reset the whole body plan, which is laid down in the earliest stages of development, without detrimental effects on the rest of the embryo. If so, then the correct explanation for why the tetrapod body plan is so taxonomically widespread involves development constraint, not just adaptation to the environment.

Spandrels, hitchhiking, vestigial traits, and developmental constraints illustrate the danger of assuming, in advance of specific evidence, that every trait of every organism has a function, or is adaptive. Gould and Lewontin were right to emphasize this, and their critique served as a useful corrective to uncritical adaptationism in biology (though arguably they caricatured their opponents' views somewhat). Contemporary evolutionists are well-aware of the methodological difficulty of determining whether a given trait has a function (in the sense of selected effect) and if so what that function is. The most careful attempts to demonstrate function involve a diverse array of evidence including cross-species comparisons, genetic analysis, experimental alteration of a trait to see its effect on fitness, and more. In the best cases, such research meets exacting scientific standards, and amounts to much more than inventing a just-so story. But it must also be admitted that in many biological quarters, there is still a predisposition in favour of finding functions, despite Gould and Lewontin's critique. Whether this represents a cognitive bias or a rational response to the prevalence of adaptive complexity in nature is a matter of opinion.

The overall moral that Gould and Lewontin draw is that it is a mistake to 'atomize' an organism into a bundle of traits and then to seek functional explanations for each trait, one at a time. Rather the organism should be treated as an 'integrated whole'. The status of this suggestion is controversial. It is quite true that the atomizing tendency is a feature of adaptationist reasoning; and certainly it is not always appropriate, for evolution cannot always alter one part of an organism without knock-on effects elsewhere. For example, if giraffes evolve longer necks, this may simultaneously reduce their running speed; so neck length and running speed cannot be independently optimized. But in practice, it has often proved possible to study the evolution of one trait considered in relative isolation from the rest of the organism. For example, biologists are able to study the evolution of the bird-of-paradise's dance without studying its dietary preferences, and vice-versa. So in some cases at least, treating an organism as if it were a bundle of separate traits is no barrier to evolutionary understanding. Gould and Lewontin were right to call attention to the atomizing tendency implicit in much evolutionary reasoning, but whether it is problematic must be assessed on a case-by-case basis.

Chapter 4
Levels of selection

The 'levels of selection' issue is one of the most fundamental in
evolutionary biology, and the subject of much controversy.
To understand the issue, consider a typical Darwinian explanation
of the sort discussed in Chapter 2, of why cheetahs have evolved
such extraordinary running speed, for example. The explanation
might go as follows:

> in the past, cheetahs varied with respect to their running speed.
> Faster cheetahs were better at catching prey than slower ones, so
> enjoyed a survival advantage, so left more offspring. Moreover,
> running speed was heritable—the offspring of faster cheetahs
> tended to be fast runners themselves. So over many generations,
> cheetahs gradually evolved to run faster and faster.

In this explanation, the process of natural selection that explains
the trait's evolution takes place at the level of the individual
organism. It is the differential survival of *individual cheetahs*—the
fact that some do better than others—that leads to the evolutionary
change. (This could also be expressed by saying that the 'unit of
selection' is the individual.) A closely related point is that the
trait in question—running fast—is explained by the advantage
that the trait confers on the individual cheetah, rather than on
some larger entity such as the whole cheetah species, for example.

So if the explanation is correct, the trait is an adaptation of individual cheetahs.

Most of the time, evolutionary biologists are concerned with selection and adaptation at the individual level, as in the cheetah example. But in theory at least, there are other possibilities. For the biological world is hierarchically structured, with smaller biological entities nested inside larger ones. The entity we call the 'individual organism' lies somewhere towards the middle of this hierarchy. Below the individual, we find entities such as cells, chromosomes, and genes; above the individual, we find entities such as families, colonies, and species. And crucially, many of these entities can in principle be subject to Darwinian evolution, for a form of reproduction applies to them. Just as organisms give rise to other organisms, so too do genes, cells, colonies, and species. The most familiar sort of natural selection, in which the selective competition is between individuals, is not the only possible sort.

Below the individual level, two types of selection process can be distinguished. The first is selection between different cell-lines within the lifespan of a multi-celled organism, known as 'somatic selection'. This occurs in the vertebrate immune system, in neuronal development, and in cancer. In this process, the entities that vary, reproduce differentially as a result, and pass on their traits to their offspring are cells. Such cell-level selection plays a role in development, but in modern organisms it rarely has long-term evolutionary consequences, for its effects are confined to an individual organism's lifespan. The second sort of sub-individual selection is different. It involves selection between the genes within a single organism. This arises because in sexual reproduction, only half of an organism's genes are transmitted to its offspring. So a form of selective competition can occur in which some genes find ways of increasing their transmission at the expense of others. Such gene-level selection does have long-term consequences; it is discussed further below.

However, it is selection above the individual level, known generically as 'group selection', that will be our starting point. At issue is whether natural selection ever acts at the group level, and thus whether there exist traits that evolved because they benefit groups rather than individuals. Perhaps surprisingly, this question is at the heart of a decades-old debate in biology that shows no sign of subsiding.

Altruism and group selection

Traditionally, biologists have invoked group selection to help solve *the puzzle of altruism* in nature. 'Altruism' in biology refers to any trait or behaviour that is costly for an individual organism, in that it reduces their Darwinian fitness, but benefits others. Altruism in this sense is quite widespread in nature. Consider a honey bee that attacks intruders to the nest by stinging them. Since the bee dies after using its stinger, its action clearly does not enhance its own survival; rather it benefits its nest-mates. Or consider the warning call that a vervet monkey gives when it sees a predator. By emitting a call, a monkey draws attention to itself, which is risky, but alerts its companions to the danger. Or consider the bacterium *Pseudomonas aeruginosa* that releases chemicals called siderophores into the environment in response to iron deficiency. This is a costly action for the bacterium itself but benefits the bacteria around it, as it frees up host-bound iron for bacterial metabolism. Examples of this sort could easily be multiplied.

On the face of it, the existence of altruism is rather puzzling. Surely evolutionary theory predicts that individuals will evolve traits that enhance their *own* biological fitness, not that of others? And yet altruistic traits do exactly the opposite. An individual that behaves altruistically pays a cost that their selfish counterparts do not, so it seems that natural selection should always favour the latter. A monkey that does not give an alarm call, or a bee that does not attack intruders, or a bacterium that does not make siderophores will enjoy a selective advantage over ones that do, for they can free-ride on the generosity of others without incurring any cost. How then can the existence of altruism be squared with basic evolutionary principles?

It is here that group selection gets its purchase. For although altruistic behaviour is costly for the individual, it may conceivably confer a benefit at the group level. That is, a group composed mostly of altruists, all working together for the common good, may enjoy a survival advantage over a group of selfish individuals who care only about themselves. So if we envisage a process of group-on-group competition, rather than individual-on-individual competition, the groups that prosper may be precisely the ones in which altruistic traits are common. In short, although altruism seems hard to explain as the outcome of individual-level selection, it could potentially evolve through selection at the group level.

Interestingly, the logic of this argument was appreciated by Darwin himself. Though mostly an individual selectionist, Darwin toyed with the idea of group selection occasionally. In *The Descent of Man* (1871), he discussed how self-sacrificial behaviour, in which individuals risk their lives to help their tribe, could have evolved in early humans. If we think in terms of individual selection, it seems that such behaviour should be selected against. As Darwin said, 'he who was ready to sacrifice his life...rather than betray his comrades, would often leave no offspring to inherit his noble nature'. However, Darwin then argued that a process of group selection, in which groups compete with other groups, could provide the explanation: 'a tribe including many members who...were always ready to give aid to each other and sacrifice themselves for the common good, would be victorious over most other tribes; and this would be natural selection'. Darwin thus understood that natural selection can operate at more than one hierarchical level, and that group-level selection can explain things that individual selection cannot.

The critique of group selection

Despite Darwin's early lead, the levels of selection issue lay dormant for many years, before re-entering biologists' consciousness in the 1960s. In large part, this was due to a book published in 1966 by George C. Williams, entitled *Adaptation and Natural Selection*. Williams's stated aim was to bring some 'discipline' to evolutionary biology. His concern was with a widespread tendency among biologists at the time to think of adaptation in terms of group rather than individual benefit, often without realizing that individual selection will not necessarily produce group-beneficial outcomes. This mistaken way of thinking, which Williams exposed very effectively, became known as 'unconscious group selectionism', or 'the good of the group fallacy'.

To illustrate Williams's point, consider an argument made by Konrad Lorenz, the Austrian biologist who won the Nobel Prize for his pioneering work on animal behaviour. Lorenz wished to explain the phenomenon of ritualized fighting in male deer. Two stags competing for females will square up to each other with their huge antlers but will rarely actually come to blows. Why does the stronger deer not kill its rival, Lorenz asked? His answer was that it would be wasteful for the species if males routinely engaged in combat with their conspecifics. Now is this a good explanation? Williams argued that it is not. It might indeed be wasteful for the species if the males fought with one another, *but that isn't the reason why they don't*. A proper explanation, Williams insisted, must show why the behaviour is advantageous for an individual deer, not for some larger collective to which the deer belongs.

To better appreciate the mistake in Lorenz's reasoning, consider an analogy. Suppose a biologist wishes to explain why earthworms have a physiology that enables them to burrow effectively through soil. Consider the following answer: 'burrowing helps to aerate the soil, which improves drainage, which benefits the local ecosystem. Therefore, natural selection led earthworms to evolve the physiology needed for burrowing.' Is this a good explanation? Arguably it is not. For although it is quite true that burrowing aerates the soil, which does indeed benefit the local ecosystem, *that is not why earthworms evolved to burrow*. A good explanation must show why the burrowing behaviour, and the physiology that allows it, is beneficial for the earthworm itself, not for some larger collective such as the ecosystem. That was Williams's key point.

In making this argument, Williams was assuming that the traits in question—such as ritualized fighting and earthworm physiology—evolved by individual-level selection. Given this assumption, his point is clearly correct. For individual selection will lead to the evolution of traits that are individually beneficial; such traits may or may not be beneficial, on aggregate, for the group or species to which the individuals belong. And even if they are, this is not a case of genuine group adaptation, but rather of what Williams called *fortuitous group benefit*. To see this distinction, consider again the cheetah's ability to run fast. This trait may indeed confer a benefit on the whole cheetah species, helping it to avoid extinction, but it is not *because* of this that cheetahs evolved to be fast runners. Running fast is an adaptation of individual cheetahs that fortuitously benefits the group, not a group adaptation. Williams argued that a failure to see this logical point had led to considerable confusion in biology.

Now Williams realized that group-level selection was a possibility; and, if it occurred, it *would* lead to genuine group adaptation, that is, to group-beneficial features that evolve because they are group-beneficial. However, he argued that group selection was unlikely to be a significant factor in evolution. For the generation time of individuals is usually shorter than that of groups, so individual selection is inherently the more powerful force. Biologists should therefore refrain from appealing to group selection unless absolutely necessary, Williams argued. This aspect of Williams's argument is more controversial than his logical point above (which is widely accepted). Many contemporary biologists agree with Williams that group selection is a weak evolutionary force, and there exist mathematical models that support this conclusion; however, a sizeable minority disagree. The jury is still out on this question.

What about altruism? As we have seen, the traditional reason for appealing to group selection was to explain how altruism can evolve, given that it is individually disadvantageous. But if group selection is not the answer, then what is? This leads us straight to *kin selection*, one of the most celebrated ideas in 20th-century evolutionary biology.

Kin selection

The basic idea of kin selection is straightforward. Consider a population containing two types of organism, altruist and selfish, who are hard-wired to behave differently. Altruists perform an action that is individually costly but benefits others, for example alerting them to danger. Selfish types do not perform this action. Let us ask: which type will be favoured by natural selection? Setting aside the group selection possibility, it seems that the selfish type will be favoured. For selfish individuals benefit from the presence of other altruists in the population but without paying the cost. Therefore natural selection should drive the altruists out.

Now this argument is correct, but it relies on a crucial assumption. It assumes that altruists *indiscriminately* assist other population members, that is, that the benefits of altruistic actions fall equally on altruists and on selfish types. This assumption might hold true, but it might not. Suppose instead that altruists are discriminating—they preferentially assist other altruists but refrain from assisting selfish types. Then the situation is quite different. For although altruists still pay a cost that their selfish counterparts do not, the benefits of altruism now fall on other altruists, rather than on all population members. In principle this could offset the cost, leading altruism to evolve.

The most obvious way that such preferential treatment can arise is if altruists help their biological relatives. For relatives are genetically similar, so if we pick any gene in an altruist, there is a greater than random chance that their relative will have a copy of that gene too. Now since we specified that the altruistic behaviour is hard-wired, it must have a genetic basis, that is, there must be a gene 'for' altruism. (This is shorthand for 'any gene that predisposes the organism to behave altruistically in specific circumstances'.) Therefore, with greater than random chance, the altruist's relatives will themselves carry the gene for altruism; so the net effect of the altruistic action may be to cause the gene for altruism to become more common in the population. In short, if altruists help their biological relatives rather than randomly chosen population members, and if altruism has a genetic basis, then natural selection can lead altruism to evolve.

This simple argument was first made explicit by the English biologist William D. Hamilton in a 1964 article. He showed that altruism will evolve when a certain condition, known as *Hamilton's rule*, is satisfied. The rule states that $rb > c$, where c is the cost paid by the altruist and b is the benefit to the recipient, both measured in terms of biological fitness. The final term, r, is the 'coefficient of relationship' between altruist and recipient, which measures how closely related they are. The higher the value of r, the greater the likelihood that the recipient of the altruistic action will also possess the gene for altruism. In a typical population, any organism has degree of relatedness $r = \frac{1}{2}$ to its offspring, $r = \frac{1}{2}$ to its full siblings, $r = \frac{1}{4}$ to its grandoffspring, and $r = \frac{1}{8}$ to its full cousins. So what Hamilton's rule tells us is that altruism will evolve so long as the cost paid by the altruist is offset by a sufficient amount of benefit to sufficiently closely related relatives.

Though Hamilton did not use the term, the selective mechanism he described was dubbed 'kin selection', and quickly became the standard explanation for how altruism can evolve. Kin selection theory predicts that organisms are more likely to behave altruistically towards relatives than non-relatives; and that the extent of the altruism will be greater, the closer the relationship. These broad predictions have been confirmed empirically. For example, in various bird species, 'helper' birds are more likely to help relatives raise their young than to help unrelated breeding pairs. In Japanese macaques, altruistic actions such as defending others from attack are preferentially directed towards close kin. And in many social insect colonies, in which workers devote themselves to protecting the colony and assisting the queen's reproductive effort, there are high levels of genetic relatedness within the colony. In general, altruism between relatives is fairly common in nature but between non-relatives it is extremely rare—just as kin selection theory predicts.

Importantly, kin selection does not require that organisms be able to discriminate relatives from non-relatives, nor to calculate coefficients of relationship. Many animals can in fact recognize their kin, often by smell, but this is not essential. What matters is that the organism should behave altruistically towards others that are *in fact* its kin. It might achieve this by having the ability to tell relatives from non-relatives; but an alternative is to use some proximal indicator of kinship. For example, if an organism behaves altruistically towards those in its immediate vicinity, then it will often end up helping its kin, since relatives tend to live near each other. Cuckoos exploit precisely this fact, free-riding on the innate tendency of birds to care for the young in their nests.

Does kin selection theory apply to humans too? This is an aspect of the broader question, discussed in Chapter 7, of whether evolutionary biology can shed light on human behaviour. Two comments will suffice for now. First, the broad prediction that organisms will treat kin and non-kin differently clearly does apply to humans. For example, most humans give considerable financial assistance to their immediate relatives, but very little to anybody else. In this respect, humans' social behaviour fits with kin selection theory. Second, unlike other species, humans regularly engage in cooperative endeavours with non-relatives. Institutions such as firms, schools, and governments, which lie at the heart of human society, rely essentially on our willingness to cooperate with others. This aspect of our social behaviour is not predicted by kin selection.

Though kin selection is a widely accepted principle in biology, certain controversies surround it. One concerns its relation to traditional group selection of the sort that Darwin discussed. Hamilton originally intended kin selection as an alternative to group selection—a way of explaining the evolution of altruism *without* invoking group-level advantage. This was also how G. C. Williams saw the matter, and it is still a popular view today. Indeed, a generation of evolutionary biologists were raised on the notion that group selection is a problematic concept while kin selection is acceptable. However, times have changed. Somewhat surprisingly, many contemporary biologists argue that kin and group selection are not rivals but are actually *equivalent*, for they represent different perspectives on the same underlying biological process.

To see the grounds for this view, consider again the honey bee's barbed stinger. A group selection explanation for why the stinger evolved would point to the survival benefit it confers on the whole colony. A kin selection explanation would point to the fact that a worker bee is closely related to the queen bee, so by Hamilton's rule the worker should be prepared to sacrifice itself if the benefit to the queen is sufficient. These explanations may sound different but in fact they are of a piece—since the colony's survival is essential to the queen's reproductive success. Indeed, when we translate the hypothesis that a bee trait 'benefits the colony' and that it 'benefits the queen' into precise mathematical terms, the two turn out to be essentially identical. Indicative of this is that models of kin and group selection share a similar mathematical structure; so a group selection explanation can often be translated into kin selection terms, and vice-versa. This supports the idea that in some cases at least, the choice between group and kin selection is a matter of convention rather than empirical fact.

The issue of how kin and group selection relate illustrates one striking aspect of the levels of selection debate in biology, which explains why it has captured the attention of philosophers. The debate involves a curious mix of empirical and conceptual questions, often intertwined. At first blush, the levels question may seem purely empirical. Given that natural selection can occur at more than one level, surely we just need to find out the level(s) at which it does occur, or has done in the past? With enough empirical data, surely the question can be straightforwardly answered? In fact, matters are not so simple. Certainly, the debate is responsible to empirical data, but there is more to it than this. For not infrequently, one finds authors who agree about the basic biological facts in a given case, but who disagree about how to identify the level(s) of selection. Such disagreements are not the 'normal' scientific ones that can be resolved by collecting data, but have a conceptual, and in some cases even an ideological, dimension.

The gene-centric view of evolution

In his book *The Selfish Gene* (1976), Richard Dawkins offered a radical new take on the levels of selection issue, in the course of defending his famous 'gene-centric' view of evolution. Dawkins argued that the real evolutionary action takes place at the level of the gene, so selection and adaptation are best thought about at this level. His starting point is the neo-Darwinist idea that all evolutionary change ultimately boils down to some genes in a population becoming more common and others declining. Therefore, Dawkins argued, we can think of each gene as engaged in a competition to bequeath as many copies of itself as possible to future generations. Organisms are simply 'vehicles' or 'survival machines' that genes have built to assist them in this task. So the phenotypic traits that we see in nature, such as birds' wings and fishes' gills, are not there because they benefit the individual organisms that display them, less still the groups to which those individuals belong. Rather, the traits are there for the benefit of the underlying genes that give rise to them! Genes 'program' their host organism to display traits, including behaviours, that will boost the genes' own chances of being passed to the next generation. The 'ultimate beneficiary' of the evolutionary process is thus the gene itself, Dawkins argued.

It is important to be clear about what Dawkins means when he talks about genes competing to leave copies in future generations. He does *not* mean that the genes within a single organism are always in competition with each other, which is not true. Though selective competition between the genes within an organism does occur, it is relatively infrequent. Most of the time, the genes in an organism cooperate, for they have a common interest in their host surviving and reproducing. Rather, Dawkins's point is that each gene is in competition with its *alleles* in the population. A gene's alleles are the variant forms of the gene that can occupy the same chromosomal locus, or slot; each allele has a slightly different DNA sequence, leading to phenotypic differences. Now any gene is necessarily engaged in a zero-sum game with its alleles: it can only increase in the population if they decline. So we can think of each gene as 'trying' to out-compete its alleles, via its effect on its host organism.

Dawkins defends his view of evolution on both logical and empirical grounds. He argues that genes have a logically privileged status vis-à-vis organisms, which uniquely equips them to play the role of beneficiary in evolution. For genes are 'replicators', that is, entities of which copies are made. Thanks to the fidelity of DNA replication, a gene in one generation is usually a near-perfect copy of the ancestral gene from which it derived. But organisms are not like this. Organisms do reproduce of course; however, sexual reproduction means that offspring contain a mixture of genetic material from their two parents. Genes in existence today are descended nearly unchanged from genes that existed hundreds of thousands of years ago; but the same is not true of individual organisms. Therefore, genes have a permanence that organisms lack, Dawkins argues; so ultimately, an organism's traits are there for the benefit of the genes.

Empirically, Dawkins argues that his gene-centric view helps make sense of numerous biological phenomena. One of these is altruism. As we have seen, altruism is hard to understand from an individual organism's perspective: why pay a cost to help others? But from a gene's perspective, the rationale is obvious. By causing its host organism to behave altruistically towards relatives, who are also likely to carry a copy of the gene, the gene is indirectly helping itself! So the central point of kin selection theory—that organisms should behave differently towards relatives and non-relatives—makes perfect sense from a gene-centric perspective. More generally, if we think of an organism's traits as strategies designed by genes to aid their propagation, as Dawkins urges, we can see immediately that there are two strategies that a gene can employ. The direct strategy, used by most genes, is to produce traits that ensure their host organism will survive and reproduce. The indirect strategy, used by a few genes, is to cause their host organism to behave altruistically towards its relatives.

Another class of phenomena that the gene-centric view can illuminate are 'outlaw' genes. These are genes that *do* spread at the expense of other genes in the same organism, that is, by gene-level selection. Recall that because of sexual reproduction, the genes in a single organism do not get transmitted to the next generation en masse. Rather, the organism produces gametes that fuse with other gametes to produce a zygote. Gametes are haploid, that is, they contain only one of each chromosome pair. This means that only half of the organism's genes make it into each gamete. Most of the time, this process works fairly, so any gene has a half chance of making it into each gamete. But some genes have devised ways of subverting the system and getting into more than their fair share of the gametes, which is clearly to their advantage. Such genes are outlaws, also known as 'selfish genetic elements' or 'ultra-selfish genes'. Outlaws often have adverse phenotypic effects on their host organism, reducing its biological fitness; but are able to spread in a population because of their transmission advantage.

Outlaws are the exceptions to the common interest that usually prevails among the genes within a single organism: they benefit themselves at the expense of the collective. Indeed, outlaws often engender genetic conflict within the organism, in which other genes evolve ways of suppressing the outlaw's actions, thus restoring harmony. Most of the time this blocking action is successful—which is just as well as otherwise organisms such as ourselves would not exist. From an individualist viewpoint, outlaws are hard to understand, for they typically harm rather than benefit the individual organism itself. Nor is there any compensating benefit for the individual's relatives, nor for any social group to which it may belong, nor for the whole species. But from a gene-centric viewpoint, outlaws' actions make perfect sense. Like all genes, they are simply looking out for themselves, and have devised a novel way of gaining an evolutionary advantage that is made possible by sexual reproduction.

How does Dawkins's gene-centric view relate to the traditional levels of selection debate, which as we have seen pitched individual against group selectionists? In his early work, Dawkins suggested that both parties to this debate were wrong: the right way to think about evolution is in terms of selection on genes, not individuals or groups. However, in his later work Dawkins adopts a different line, arguing that his gene-centric theory is not meant as an empirical alternative to ordinary individual selection, nor to group selection. Rather, it is simply a different perspective on evolution that is heuristically valuable in certain contexts. We can either view evolution in the standard Darwinian way, as involving selection between individual organisms (or possibly groups in some cases); alternatively, we can switch perspective and view the process in terms of selection between genes. There is no fact of the matter as to which is right, Dawkins suggests.

This idea of alternative perspectives is compelling, but it sits uneasily with the emphasis that Dawkins places on outlaw genes. For as we have seen, outlaws do not benefit the individual organism, nor its group. So where outlaws are concerned, the freedom to switch perspectives does not seem to exist. How should this tension be resolved? The best way is to sharply distinguish the *process* of gene-level selection from a gene-centric *perspective* on selection processes that occur at other levels. (This distinction is there in Dawkins's work, but not made fully explicit.) The process of gene-level selection refers to selection between the genes within a single organism, as in the case of outlaws. So gene-level selection is a distinct level of selection of its own, to be contrasted with individual- and group-level selection. Since most genes are not outlaws, the process of gene-level selection is relatively uncommon. However, it is always possible to adopt a genic perspective on selection processes that occur at other levels, such as the individual level. For the net effect of individual selection is that some genes spread at the expense of their alleles. So, if we wish, we can regard traits that evolve by individual selection, such as the cheetah's running speed, as adaptations for the benefit of the genes.

The major transitions in evolution

In the past twenty-five years, the levels of selection debate has been revitalized thanks to a new body of work on the 'major transitions in evolution', set in train by John Maynard Smith and Eörs Szathmáry. These transitions occur when a number of smaller biological units, originally capable of surviving and reproducing alone, become aggregated into a single larger unit, generating a new level in the biological hierarchy. Such transitions are thought to have occurred repeatedly in the history of life. The very first transition was probably from solitary replicators (made of RNA) to networks of replicators enclosed within compartments. Later transitions were from independent genes to chromosomes consisting of a number of genes physically linked together; from prokaryotic (or bacteria-like) cells to eukaryotic cells containing organelles (such as mitochondria and plasmids); from single-celled eukaryotes (such as amoeba) to multi-celled organisms (such as most animals and plants); and from solitary organisms (such as wasps) to integrated colonies (such as honey bees). In each case, the transition involved free-living biological units evolving to becoming parts of a larger whole. The challenge is to understand this in Darwinian terms. Why was it advantageous for the smaller units to sacrifice their individuality, cooperate with one another, and form themselves into a corporate body?

In thinking about the major transitions, we immediately run up against the levels of selection issue. Take for example the transition to multi-cellularity. The earliest multi-celled organism was probably a loose aggregate of a few hundred cells, more like a colony than a 'real' organism. Over time it evolved into a highly cohesive unit, containing trillions of cells with specialized tasks, all working for the common good. But why did selection between cells not disrupt the integrity of the emerging multi-celled organism? For as we know, selection at one level need not have beneficial effects at higher levels. Various answers to this puzzle have been suggested. One is to invoke kin selection: perhaps the cells within the emerging multi-celled aggregate were close relatives, or even clones, and that is why they cooperated? This could occur if the organism's lifecycle evolved to pass through a single-celled stage—which is exactly what we see in most modern plants and animals, which develop from a single zygote. This ensures that their constituent cells are clonally related, or nearly enough. (In terms of Hamilton's rule, clonality means that $r = 1$.)

On this theory, the explanation of why individual cells gave up their free-living lifestyle and evolved into parts of a larger unit is that by doing so they could help their genetic relatives.

The work on major transitions has interesting conceptual implications. For one, it suggests that the traditional formulation of the levels of selection issue was somewhat inadequate. The traditional formulation, which we followed above, takes the existence of the biological hierarchy for granted, as if it were simply a God-given fact about the world. But of course, the biological hierarchy is itself the product of evolution—entities further up the hierarchy, such as multi-celled organisms, were obviously not there at the dawn of life on earth. The same is true of cells and chromosomes. So ideally, we would like an evolutionary theory which explains how the biological hierarchy came into existence, rather than treating it as a given. From a major transitions perspective, it is not enough to consider selection and adaptation at pre-existing hierarchical levels; we also need to understand how the levels in the hierarchy evolved in the first place.

This lends the levels of selection debate a renewed sense of urgency. Some biologists were inclined to discuss the traditional debate as a storm in a teacup—arguing that, in practice, selection on individual organisms is the only important evolutionary force, whatever the other theoretical possibilities. But in the light of the major transitions, this attitude is hard to defend. What we call an 'individual organism' is itself a highly cooperative group of cells; and a single eukaryotic cell is itself a group of a sort, since it was formed by the merger of two prokaryotic cells that used to have an independent existence. In short, the 'individual organisms' of today did not always exist and, ironically enough, they evolved into cohesive individuals through a process of group-level selection (where 'group' means group of cells). Therefore, selection at levels other than that of the individual organism must have occurred in the past, whether or not it still occurs today.

From this expanded point of the view, the argument that individual selection is 'all that matters in practice' is clearly unsustainable.

A second interesting implication of the major transitions is that in a sense, all of life on earth is social. That is, all biological organisms, including you and I, are in effect complex social groups, built up out of smaller units, such as cells and organelles, that are the descendants of free-living ancestors. This is quite a striking thought. For while it is obvious that an insect colony or a baboon troop is a social group, it is far less obvious, but nonetheless true, that every multi-celled organism, and indeed every eukaryotic cell, is also a social group. Thinking of life as social in this way forces us to rethink what an 'individual' is, and to ask whether the distinction between 'individuals' and 'groups' might be context-relative. (That is, perhaps a single entity can count as an individual in some contexts but as a group in others.) And it helps us to understand why the key concepts of social evolution theory, such as group selection, kin selection, and Hamilton's rule, that were originally devised to explain specific animal behaviours, should have a much broader domain of application.

Chapter 5
Species and classification

An important part of scientific enquiry involves classifying the objects under study into distinct kinds, or types. Physicists classify fundamental particles as baryons, leptons, or mesons depending on their mass. Astronomers classify galaxies as elliptical, spiral, or irregular depending on their visual appearance. Geologists classify rocks as igneous, sedimentary, or metamorphic depending on how they were formed. Part of the point of classification is to convey information. If you find a rock and a geologist tells you that it is igneous, this tells you a lot about its likely behaviour. So a good classification scheme should group together objects that are alike in important respects, and which are thus expected to behave similarly.

Classification in science raises a deep philosophical issue. For notice that, in principle, all objects can be classified in more than one way. For example, fundamental particles can be classified by their spin instead of their mass, which yields a division into two types: bosons and fermions. So how should we choose between the alternative ways of classifying? Is there a 'correct' way to classify the objects in a given domain, or are all classification schemes ultimately arbitrary? This question arises quite generally, but our focus here is on how it plays out in relation to biological classification, or taxonomy.

Biologists traditionally classify organisms using the *Linnaean system*, named after the 18th-century Swedish naturalist Linnaeus. The basics of the Linnaean system are straightforward. First, each individual organism is assigned to a *species*. Next, each species is assigned to a *genus*, each genus to a *family*, each family to an *order*, each order to a *class*, each class to a *phylum*, and each phylum to a *kingdom*. The species is thus the base taxonomic unit; while genera, families, orders, etc. are known as 'higher taxa'. To take an example, my pet cat belongs to the species *Felis catus*, which along with a handful of other small cat species makes up the *Felis* genus. This genus itself belongs to the family Felidae, the order Carnivora, the class Mammalia, the phylum Chordata, and the kingdom Animalia. Note that a species' Latin name indicates the genus to which it belongs, but no more.

A notable feature of the Linnaean system is its hierarchical structure. A number of species make up a single genus, a number of genera a single family, and so on. So as we move upwards, we find fewer taxa at each rank. At the bottom there are literally millions of species, but at the top there are just a handful of kingdoms: animals, plants, fungi, protists, archaea, and bacteria. Not every classification system in science is hierarchical in this way. For example, the chemical elements are grouped according to the vertical column in the periodic table in which they lie; but these groups are not then nested into further groups as in the Linnaean system. One interesting question, to which we shall return, is *why* biological classification should be hierarchical.

The Linnaean system served biologists well for years, and elements of it are still used today. In some ways this is surprising, since the underlying scientific worldview has changed greatly. Linnaeus belonged to the pre-Darwinian era and was a devout Christian who accepted the biblical story of creation. He regarded his classification system as an attempt to discover the objective, eternal divisions between living organisms that God had created.

For Linnaeus, the idea that contemporary species are descended from common ancestors would have been entirely foreign.

To understand how the Linnaean system could survive the transition from a creationist to an evolutionary worldview, recall that evolution is a very slow process. So even though all life forms have descended from a common ancestor, this is compatible with the existence of discontinuities among the organisms that exist today. And such discontinuities can certainly be found. Organisms seem to cluster into discrete types, many of which are easily recognizable. A hamster, after all, seems objectively different from a mouse or squirrel—even though, if we traced their ancestry back far enough, we would arrive at a life form that does not fit easily into any of these three taxa. So the fact of evolution does not automatically undermine the attempt to find an objective way of classifying contemporary organisms. Indeed, many of the species, and some of the higher taxa, that Linnaeus identified are still recognized by modern biologists.

Having said that, the rise of evolutionary biology did eventually lead to fundamental changes in both the theory and practice of biological classification. Indeed, the sub-discipline of taxonomy, or 'systematics' as it is often known today, came into its own in the 20th century, prompted by the need for a clearer articulation of the principles to be used in classification. Interestingly, biologists have not always agreed on what these principles are, in part due to underlying philosophical differences. Beginning in the 1970s, this led to a prolonged debate about biological classification that continues to this day, to which philosophers have made substantial contributions.

The problem of biological classification divides into two. First, how should organisms be assigned to species? Second, once this has been done, how should the species then be organized into higher taxa? Though related, these questions raise somewhat different issues, so merit separate treatment.

The species problem

Biologists often discuss what they call 'the species problem'. This refers to the problem of giving a precise definition of what a species is. Somewhat surprisingly, there is no agreement on this matter. Competing definitions of species, or 'species concepts' as they are called, abound. This disagreement has practical consequences. For it means that not infrequently, biologists will disagree about how many species a particular taxon actually contains. For example, consider the Bovidae family of cloven-hoofed ruminant mammals (which includes bison, antelope, sheep, and cattle). Traditionally, the Bovidae was thought to contain 143 extant species. However, in 2011 a group of experts recommended that 279 bovid species be recognized—not because of any new empirical discoveries, but because they favoured a different species concept. Other experts rejected the recommendation as unwarranted taxonomic inflation. Such disagreements are symptomatic of the species problem not yet having been resolved.

Non-biologists are often surprised to learn that there is a species problem. For the word 'species' is part of everyday English, where it has a meaning fairly similar to its meaning in professional biology. Moreover, people without any biological training can often make correct judgements about species membership. A 3-year-old child can tell that two animals in the park are both dogs, even if they are of different breeds; and a biologist will confirm that the child is correct—the animals do indeed belong to the same species, namely *Canis familiaris*. Thus it is natural to think that there is a straightforward matter of fact about which species any given organism belongs to. Most non-biologists appear to accept this view without question.

This common-sense viewpoint dovetails with the philosophical doctrine of 'natural kinds', variants of which have been popular since Aristotle. This doctrine holds that there are ways of grouping objects into kinds that are natural in the sense of corresponding to divisions that really exist in the world, rather than reflecting human interests. Chemical elements and compounds are paradigm natural kinds. Consider, for example, all the lumps of pure gold in the universe. These lumps are of the kind 'gold' because they are alike in a fundamental respect: their constituent atoms have atomic number 79. By contrast, a lump of fool's gold (iron pyrite) does not belong in that kind, despite being similar to gold in some respects, for it is a compound formed by atoms of a different sort (iron and sulphur). Similarly, it has often been suggested that species are the natural kinds of biology.

In fact, however, biological species are rather unlike the natural kinds we find in chemistry and physics. For a kind such as gold, we can point to a single property—having atomic number 79—which is necessary and sufficient for belonging to the kind, and which thus constitutes the 'essence' of gold. But for a biological species, this is generally not possible. The reason is simple: in every species, we find considerable *variation* among its constituent organisms. Mutation continually throws up new genetic variants, and sexual reproduction continually 'shuffles' genes around, resulting in extensive genetic differences between the organisms within a single species. Furthermore, the genetic make-up of any species changes over time, as it evolves. So unlike for gold, we cannot easily point to any property which is necessary and sufficient for belonging to, say, *Canis familiaris*.

This is not to deny that in practice, biologists can often tell which species an organism belongs to by sequencing its DNA. Indeed such 'DNA barcoding', as it is known, is often a fairly reliable way to determine species membership. For it is usually possible to find a DNA sequence that shows relatively little within-species variation, but does vary between species. However, DNA barcoding does not always work, and even when it does, it does not show that membership in a species is determined by some fixed 'genetic essence', as the traditional doctrine of natural kinds requires.

In the *Origin of Species*, Darwin offered an interesting perspective on the species problem. He observed that biologists often recognize groupings of organisms below the species level—they talk of 'varieties', 'breeds', and 'races'. (Today the term 'subspecies' is often used too.) The need for such terms arises because there are often recognizable clusters within a species, that are not quite different enough to count as separate species. But where do we draw the line? What determines when we have two varieties of a single species as opposed to two species? Darwin argued that there is no sharp line to be drawn. He wrote:

> I look at the term species as one arbitrarily given, for the sake of convenience, to a set of individuals closely resembling each other, and that it does not essentially differ from the term variety, which is given to less distinct and more fluctuating forms.

Darwin's suggestion that there is an element of arbitrariness in what counts as a species is somewhat surprising, given the title of his book. But is Darwin right about this? In the second half of the 20th century, many biologists became convinced that species are in fact real units in nature, not arbitrary groupings, on the grounds that they are *reproductively isolated*. This means that the organisms within a species can interbreed with each other to produce fertile offspring, but not with those of other species. Defining species in terms of reproductive isolation was championed by Ernst Mayr, whose 'biological species concept' (BSC) is perhaps the best-known attempt to solve the species problem.

The biological species concept

The key idea behind the BSC is that the similarities and discontinuities among living organisms that motivate the attempt to divide them into species in the first place arise because of *restricted gene flow*. To understand this, consider two closely related species, for example the common chimpanzee (*Pan troglodytes*) and the bonobo (*Pan paniscus*). These two species share a recent common ancestor, but they are clearly distinct in their appearance and behaviour. Now the differences between chimps and bonobos arose initially, and persist today, because the two groups do not interbreed. As a result, mutant genes that arise in bonobos cannot 'flow' into the chimpanzee gene pool, nor vice-versa. This is what allowed the two groups to diverge genetically in the first place, and explains why they retain their distinct identities today. If chimps and bonobos were to readily interbreed, the differences between them would quickly disappear.

The BSC is an attempt to generalize this moral into an explicit definition of a species. In Mayr's words, 'species are groups of actually or potentially interbreeding natural populations that are reproductively isolated from other such groups'. This is an attractive definition, and the species it picks out often correspond closely to the species that earlier biologists had recognized. The BSC thus supports the idea that the boundaries between species are real rather than conventional. Also, the BSC implies that there *is* a principled distinction between varieties and species, contrary to what Darwin thought. Consider, for example, the European and American golden eagles. The BSC counts these as two varieties of a single species, not separate species, since they can in principle interbreed and produce viable offspring (even if they do so rarely). By contrast, the spotted eagle and the golden eagle count as separate species, since their members cannot interbreed.

The BSC marked an important advance in our understanding of species, and it is still widely used today. However, it has its limitations. For one, it only applies to sexually reproducing organisms; but many living organisms reproduce asexually, including most microbes, some plants and fungi, and a few animals. For such organisms, the BSC offers no insight into what a species is, so at best it is a partial solution to the species problem.

Moreover, reproductive isolation is not always a hard and fast matter, but comes in degrees. Closely related species that live in adjacent locations often have 'hybrid zones' where their ranges meet; in these zones, a limited amount of hybridization takes place, in which fertile offspring are produced at least some of the time, but the two species retain their distinct identities. Hybrid zones often arise when one species is in the process of splitting into two, but they can persist for a long time. Among plants, in particular, hybridization between organisms that belong to clearly distinct species is quite common. So if applied literally, the BSC would give us the 'wrong' answer in these cases.

Another problem for the BSC arises from *ring species*. This term describes a species composed of a number of populations arranged geographically in a ring, where each population can interbreed with its immediate neighbour, but the populations at either end of the ring cannot. For example, the salamander species *Ensatina eschscholtzii* is made up of a number of distinct populations, arranged in a ring-like shape around the mountains of the Central Valley in California (Figure 2). Each population can interbreed with a neighbouring one, but the population at the western end of the ring (X) cannot interbreed with the one at the eastern end (Y). This constitutes a kind of paradox for the BSC. To see why, let us ask whether populations X and Y belong to the same species or not? Since they cannot interbreed, the answer should be 'no'. However, X and A are conspecific, since they can interbreed; and similarly for A and B, B and C, C and D, and so on. Thus, by logic, we can infer that populations X and Y are conspecific after all! So attempting to define species by the interbreeding criterion leads to paradox.

Though ring species pose an interesting logical problem, they are not fatal to the BSC, for two reasons. First, they are relatively infrequent in nature. Second, they are usually thought to represent a stage in the process of speciation, that is, the two populations at either ends of the ring are incipient new species. In general,

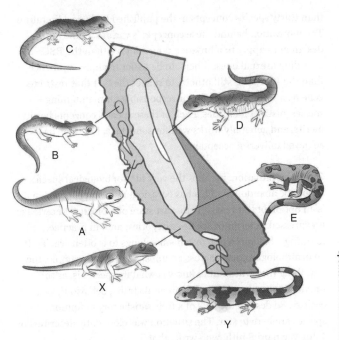

2. Local populations of _Ensatina eschscholtzii_ form a ring in which the westernmost and easternmost populations cannot interbreed.

it takes thousands of generations for one species to split into two reproductively isolated groups; so transitional forms, and populations whose status is unclear, are only to be expected. From this perspective, the existence of ring species does not point to a shortcoming in the BSC that could be remedied by finding a better definition, but rather reflects the fact that perfectly neat species boundaries will not always exist, given how evolution works.

The shortcomings of the BSC have motivated the development of various alternatives. These include the 'ecological species concept', the 'phylogenetic species concept', the 'morphological species concept', and more. Indeed, a recent survey finds no fewer than thirty species concepts in the published biological literature. The motivation behind these concepts is various. Some are designed to apply to a broader range of taxa than the BSC, including asexual ones; others, to be easier to apply in practice than the BSC; and still others, to reflect the fact that restricted gene flow is not the only factor responsible for maintaining species' distinct identities. Each of these species concepts has its merits, and generally works well for some taxa, but none has achieved universal acceptance.

Given this situation, perhaps the very idea of biological species should be abandoned? This has occasionally been proposed, but it is an option of last resort. For a vast store of biological knowledge is expressed in terms of species categories; and, in practice, knowing what species an organism belongs to is often crucial. If an ornithologist comes across an unusual bird, for example, the first thing they will want to know is what its species is, as this provides valuable information about its traits, behaviour, and ecology. So despite the lack of a fully satisfactory definition, species are here to stay. The situation was eloquently described by John Maynard Smith, who wrote that

> any attempt to divide all living organisms, past and present, into sharply defined groups between which no intermediates exist, is foredoomed to failure. The taxonomist is faced with a contradiction between the practical necessity and the theoretical impossibility of his task.

So biologists continue to treat species as if they were sharply defined groups, in the knowledge that this is only an approximation to reality.

Species as individuals

In the late 1970s, an intriguing diagnosis of the species problem was given by the biologist Michael Ghiselin and the philosopher David Hull. They argued that the problem as traditionally formulated rested on a mistaken assumption, namely that a biological species is a kind, or type, of thing. Instead, they argued that a species is a complex individual; that is, a *particular* thing. To understand this, consider an ordinary biological individual such as Red Rum (the British racing horse from the 1970s). Red Rum was born at a particular time and place, and had a finite lifespan, dying in 1995. Similarly, a biological species comes into existence at a particular time and place, when a speciation event takes place, and persists for a certain duration until it goes extinct. By contrast, a genuine kind is unrestricted in time and space. Consider the kind gold. A piece of matter anywhere in the universe counts as gold, irrespective of its origin, so long as its constituent atoms have atomic number 79. So in principle, all the gold in the universe could be destroyed and then years later some more could be synthesized. But species are not like this, Ghiselin and Hull argued. Once a species goes extinct it cannot come back into existence as a matter of logic, any more than you or I can survive our deaths.

Ghiselin and Hull did not defend any particular species concept, nor did they seek a criterion for actually delimiting species in the field. Rather, they sought to re-orient the species discussion by rethinking its philosophical basis. Also, Ghiselin and Hull did not deny that the species *category* is a kind. The species category contains as its members *Homo sapiens*, *Canis familiaris*, *Felis catus*, and every other species. Rather, their claim was that each species *taxon* is an individual rather than a kind. This means that the relation between an organism and its species is that of part to whole, not member to kind. To understand this, consider the relation between a particular cell in Red Rum's body and Red Rum himself. This is a part–whole relation: the cell is a part of Red Rum, not a member of Red Rum. Similarly, the relation between Red Rum and *Equus ferus caballus* (the domesticated horse species) is part–whole, according to Ghiselin and Hull.

The species-as-individuals idea appears strange at first sight. For species are unlike 'ordinary' biological individuals in that their constituent organisms are not joined together. However, this difference is fairly superficial. The ants in an ant colony are not joined together either, but we are quite happy to regard the whole colony as an individual. Moreover, treating species as individuals has distinct advantages. In particular, it helps to reconcile biologists' long-standing belief that at least some species are real units in nature, not arbitrary groupings, with the fact that they do not have 'genetic essences' and that their constituent organisms vary. If a species were a natural kind, we should expect there to be a necessary and sufficient condition for belonging to the species—as there is for gold, for example. Failure to find such a condition should make us doubt whether the kind marks a real division in nature. However if a species is a complex individual, whose constituent organisms are its parts rather than its members, there is no such expectation. For the parts that comprise a whole do not do so in virtue of all possessing some essential property, or satisfying some necessary and sufficient membership condition. The absence of such a property is thus perfectly compatible with the reality of the whole.

To see this point, consider a non-biological example. The desk in my office has a glass top and steel legs. So the two parts of my desk—the top and the legs—are intrinsically unalike. But that does not prevent them from being parts of a single thing. Moreover, even if they *were* intrinsically alike, for example if both were made of steel, it would not be in virtue of this that they constitute parts of the same table. The same holds true of biological part–whole relations. The cells in my body are genetically very similar, but it is not in virtue of this that they are part of me. A mutant cell in my liver is still part of me, and a cell in my identical twin is not part of me, despite being genetically identical to my cells. The same is true of organisms and species, on the species-as-individuals view. The extensive genetic variation that we find among the organisms in a species in no way compromises the reality of the species, so long as we regard the species as an individual rather than a kind.

Hull argued that the species-as-individuals idea has striking philosophical implications. One is that we should not expect to discover scientific laws that apply to all and only the organisms in a given species. A long tradition in philosophy of science, discussed in Chapter 1, sees science as the search for exceptionless laws of the form 'for all objects x, if x is of kind F then x is of kind G'. (All electrons are negatively charged; all metals conduct electricity; all planets orbit the sun.) Now if we assume that a species is a natural kind, so is a candidate for 'F' in the above schema, we might expect biologists to discover attributes (candidates for 'G') that apply to all and only the organisms in a species. But if species are complex individuals whose constituent organisms are their parts, there is no such expectation, and the species is not a candidate for 'F' in the first place. The fact that exceptionless generalizations about the members of a biological species are typically hard to find thus lends indirect support to the species-as-individuals thesis.

A second philosophical implication concerns the age-old debate about human nature. Scholars down the ages have assumed that there is such a thing as human nature—some essential property that defines 'what it is to be human'. (Various candidates for what this property is have been offered.) Belief in human nature seems reasonable, since humans are clearly different from chimpanzees, our nearest living relatives. But Hull argued that since *Homo sapiens* is an individual not a kind, there is no property that defines the essence of humanity. This is not to deny that humans have attributes that chimps lack, of course, but rather to reject the claim that it is in virtue of possessing some particular set of attributes that someone counts as a human. If Hull is right, then much of the traditional debate about human nature rests on a false assumption.

Phylogenetic systematics

Once organisms have been satisfactorily assigned to species, the next step in biological classification is to organize species into higher taxa. What principles should be used to do this? The standard answer is given by *phylogenetic systematics*, also known as *cladistics*, which is the dominant taxonomic methodology today. The key idea of phylogenetic systematics is that classification should 'reflect evolutionary history', that is, species should be grouped according to how closely related they are. More precisely, all taxonomic groups above the species level, be they genera, families, orders, or whatever, are required to be *monophyletic*, according to phylogenetic systematics.

A monophyletic group, or clade, is one which contains *all and only the descendants of a single ancestral species*. Put differently, the species in a monophyletic group must share a common ancestor which is *not* ancestral to any species outside the group. Monophyletic groups come in various sizes. At one extreme, all species that have ever existed form a monophyletic group, presuming life on earth only originated once. At the other extreme, there can be monophyletic groups of just two species—if they are the only descendants of a common ancestor. Groups that are not monophyletic should not be recognized in biological classification, according to phylogenetic systematics, irrespective of how similar their members may be, for they are 'artificial' rather than 'real' groupings.

To understand the concept of monophyly, consider Figure 3, which is a phylogenetic tree depicting the pattern of ancestry among primates. The forks in the tree depict speciation events, when an ancestral lineage split into two. The primates form a monophyletic group, as they comprise all and only the descendants of a common ancestor (which lived some sixty-three million years ago), depicted at the base of the figure.

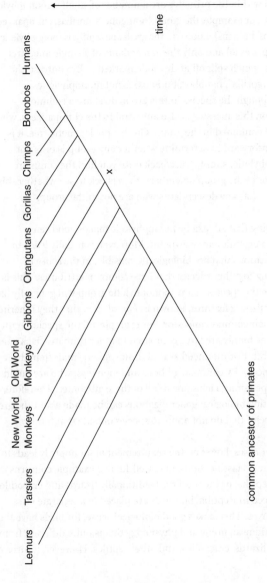

3. Phylogeny of the major primate groups.

Nested within the primates are a number of smaller monophyletic groups. For example, the group {orangutans, gorillas, chimpanzees, bonobos, humans}, known as the great ape family, is monophyletic: it comprises all and only the descendants of a single ancestral species, which split off at the node marked x. By contrast, the group {gorillas, bonobos, humans} is not monophyletic. For although the species in this group do share a common ancestor, this ancestor is also ancestral to the chimpanzees, who are not contained in the group. Given a phylogenetic tree, it is straightforward to determine whether any group is or is not monophyletic: simply trace back until you find the common ancestor of the group's members, then check to see whether this ancestor has any descendants who are not in the group.

Requiring that all taxa be monophyletic makes good sense from an evolutionary viewpoint. Moreover, it usually leads to classifications that are biologically sensible, in the sense of grouping together species that have common attributes. This is because the species in any monophyletic group will generally have distinguishing features, known as *homologies*, that they inherited from their common ancestor. For example, all the species in the great ape family are large compared to other primates, lack a tail, and exhibit pronounced sexual dimorphism (male/female differences). By contrast, gibbons and monkeys, who are not in the great ape family, lack some or all of these attributes. Therefore, interesting biological generalizations can be made about the great ape species that do not apply to species outside that family.

In other cases, however, the requirement of monophyly leads to 'unnatural' classifications. One well-known example concerns the class Reptilia, or the reptiles. Traditionally lizards and crocodiles are placed in Reptilia, but birds are placed in a separate class called Aves. This makes good biological sense, for birds have their own unique anatomy and physiology that is quite different from that of lizards, crocodiles, and other reptiles. However, it turns out

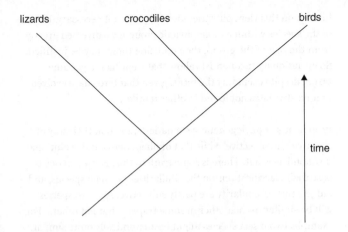

lizards   crocodiles   birds

time

4. Reptilia is not monophyletic, since it excludes birds.

that Reptilia is not monophyletic, as Figure 4 shows. The common ancestor of the lizards and the crocodiles is also an ancestor of the birds; so placing lizards and crocodiles together in a group that excludes birds violates the requirement of monophyly. Phylogenetic systematists therefore recommend that traditional taxonomic practice be abandoned: we should not recognize Reptilia at all, for it is not a real taxon. It is simply a mistake to think that there are any such things as reptiles, they argue.

The Reptilia example, and others like it, were the focus of an intense debate between proponents of phylogenetic systematics and two rival taxonomic schools in the 1960s and 1970s. According to the *phenetic* school, classification should have nothing do with evolutionary history, but should instead be based on observed similarities between species. The aim should be to group together species that are similar, irrespective of common ancestry. So phenetics is diametrically opposed to phylogenetic systematics. The third school, *evolutionary taxonomy*, attempts to steer a middle course. Classification should reflect evolutionary history, on this view, but strict monophyly is not necessary. If some of the species within a monophyletic group have diverged greatly from the rest of the group, then omitting them may be justified. So evolutionary taxonomists allow that Reptilia is a genuine taxon, despite excluding the birds, given that birds have evolved unique attributes not found in other reptiles.

Proponents of phylogenetic systematics argue that their way of classifying is 'objective' while that of pheneticists and evolutionary taxonomists is not. There is some truth to this. For pheneticists base their classifications on the similarities between species, and judgements of similarity are partly subjective. Any two species will be similar to each other in some respects but not others. For example, two insect species might be anatomically quite similar, but have different feeding habits. So which 'respects' do we single out, in order to make judgements of similarity? The same problem afflicts evolutionary taxonomists. For they need to make decisions about the extent of divergence between groups—such as that birds have diverged a long way from the usual reptilian taxa. Such decisions rest partly on 'sound biological judgement', so cannot be fully objective. By contrast, the criterion of monophyly is perfectly sharp: a given group either is or is not monophyletic (though we may not know which). This is one reason why phylogenetic systematics is the dominant approach to biological classification today.

In the early days of phylogenetic systematics, the objection was often voiced that it introduced too much uncertainty into classification. It is all very well to require that taxa be monophyletic, but this is of limited use unless we can tell whether a given group is monophyletic or not. This in turn requires that we know what the true phylogenetic tree looks like, but we can only know this indirectly, via inference. So on the phylogenetic approach, whenever we classify we are implicitly making a hypothesis about the phylogenetic relations between the species in question—and the hypothesis may turn out to be wrong. Biological classification thus becomes a work in progress, subject to revision as our knowledge of phylogeny improves.

This is a valid concern, and it is certainly true that the monophyly requirement has necessitated much taxonomic revision. But in the past twenty-five years, biologists have become much better at inferring phylogenetic trees. For thanks to molecular biology, they now have a rich new source of data: the DNA sequences of organisms. Traditionally, systematics had to rely on morphological traits—such as skull shape and skeletal structure—in order to reconstruct the phylogenetic relations among species. But DNA sequences offer a far more reliable way of determining such relations, in part because of their specificity, and in part because of the sheer number of them. Moreover, sophisticated statistical methods have been developed to analyse the molecular data, which enables biologists to be much more certain about phylogenetic relations than they once were. To take one of many examples, molecular data have helped to resolve an old debate about whether humans share a more recent common ancestor with chimpanzees or gorillas. (The answer is chimpanzees.) So the objection that basing classification on phylogeny engenders too much taxonomic uncertainty carries much less weight than before.

Finally, how does phylogenetic systematics relate to the traditional Linnaean classification scheme? It vindicates some aspects of the Linnaean scheme, such as the hierarchical nature of classification. As is clear from Figure 3, monophyletic groups are always nested inside each other, with no overlapping; so if the requirement of monophyly is followed, the resulting classification will automatically have a hierarchical structure. However other aspects of the Linnaean scheme sit less well with phylogenetic systematics. Recall that each taxon in the Linnaean scheme has a particular rank: genus, family, order, and so on. From the perspective of phylogenetic systematics, these ranks have no intrinsic meaning—there is no principled basis for saying whether any given monophyletic group counts as a genus or a family, for example. For this reason, some biologists argue that the Linnaean ranks should simply be dispensed with altogether; however, this remains a minority viewpoint.

Chapter 6
Genes

Of all the entities that biologists talk about, the *gene* is perhaps one of the best-known. The notion that the genes within an organism, including a human, are responsible for many of the organism's observable traits, such as their skin colour, is firmly enshrined in the popular consciousness. Advances in the field of genetics frequently make newspaper headlines, particularly when they are relevant to human health. Indeed, hardly a week goes by without medical researchers announcing the discovery of a gene that is implicated in producing a disease. A famous example is the discovery of the gene that causes Huntington's disease, a lethal neurodegenerative condition, by a team at MIT in 1993. And genes were at the heart of one of biology's largest collaborative research efforts, the Human Genome Project, which ran from 1991 to 2003. It is not for nothing that the historian of biology Evelyn Fox Keller has described the 20th century as 'the century of the gene'.

Let us start with a deceptively simple question: what exactly *is* a gene? Somewhat surprisingly, given the prominent position of genetics within the biosciences, there is no satisfactory one-line answer to this question. This is not as paradoxical as it sounds, for many important scientific concepts cannot be given a perfectly precise definition—think, for example, of the species concept discussed in Chapter 5. But in the case of the gene, the reasons why a precise definition is elusive are particularly interesting, and raise a number of philosophical subtleties. To understand them, we need to delve briefly into the history of genetics.

Mendelian and classical genetics

The origins of genetics lie in the work of Gregor Mendel, a Czech monk, in the 1860s. Mendel conducted breeding experiments with pea-plants, in order to study how particular plant traits were transmitted across generations. One such trait was the shape of the plants' peas, which could be either round or wrinkled. In a famous experiment, Mendel began with two pure-breeding plant lines, one with round peas and the other wrinkled. He then crossed the two lines to produce a generation of hybrid plants. These F1 hybrids, as they are known (for 'first filial generation'), all had round peas; so the wrinkled trait seemed to have disappeared from the population (Figure 5). Next, Mendel crossed the F1 hybrids with each other, to produce the F2 generation. He found that ¾ of the F2 plants had round peas, while ¼ of them had wrinkled peas. So the wrinkled trait had made a mysterious comeback. It was as if the capacity for producing wrinkled peas had lain dormant in the F1 plants, only to be somehow re-activated in the F2s. Mendel found that the same held true for other plant traits that have two variants. In each case, the F1 hybrids were all alike, while ¼ of the F2s had the variant that was absent from the F1 generation.

Mendel offered a simple but ingenious explanation of this finding. He suggested that a plant's pea shape is determined by a pair of 'factors'. A plant inherits one factor from each of its parents. The factors are of two types: **R** (for round) and **W** (for wrinkled). So there are three possible types of plant: **RR**, **RW**, and **WW**. Now an **RR** plant will have round peas, while a **WW** plant will have wrinkled peas. What about an **RW** plant? Mendel suggested that it will have round peas, since the **R** factor is 'dominant' and the **W** factor is 'recessive'. This implies that **RR** and **RW** plants

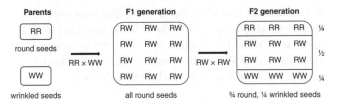

Parents	F1 generation	F2 generation

Parents

RR
round seeds

WW
wrinkled seeds

RR × WW

F1 generation

RW	RW	RW
RW	RW	RW
RW	RW	RW
RW	RW	RW

all round seeds

RW × RW

F2 generation

RR	RR	RR	¼
RW	RW	RW	
RW	RW	RW	½
WW	WW	WW	¼

¾ round, ¼ wrinkled seeds

5. In Mendel's experiment, all the pea plants in the F1 generation had round seeds, but the F2 generation contained both round and wrinkled-seeded plants in the proportion 3:1.

will be indistinguishable. Finally, Mendel suggested that the pair of factors in a single plant 'segregate', so the plant passes only one of the pair, chosen at random, to each of its offspring. This is known as *the law of segregation*. With these ingredients, Mendel was able to explain his experimental results. His two initial pure-breeding lines were of type **RR** and **WW** respectively. So the F1 hybrids were all of type **RW**, and thus all had round peas. Now the F2 plants were formed by **RW** x **RW** crosses. So by the law of segregation, we should expect all three types to be present among the F2s, in the proportions ¼ **RR**, ½ **RW**, and ¼ **WW**. Given that the **RR** and **RW** types both have round seeds, it follows that ¾ of the F2 plants will have round seeds and ¼ wrinkled, which was precisely Mendel's experimental finding.

Mendel introduced a second law designed to explain a more complicated set of experimental findings about the inheritance of multiple traits. Consider two traits, such as pea shape and flower colour, each with two variants (wrinkled vs round, and yellow vs green). So there are four possible trait combinations: wrinkled & yellow, wrinkled & green, round & yellow, round & green. Again, Mendel studied the proportions of these trait combinations in the progeny of crosses. He found that the data could be explained by the hypothesis that each trait is controlled by a pair of factors, as per his first law, which segregate *independently* of each other. To see what this means, consider a plant of type **RW/YG**, that is with one **R** and one **W** factor controlling seed shape, and one **Y** and one **G** factor controlling flower colour. The plant will transmit to each offspring either the **R** or the **W** factor, and either the **Y** or the **G** factor. Mendel's second law says that whether a given offspring receives **R** or **W** from its parent has no bearing on whether it receives **Y** or **G**. This is known as *the law of independent assortment*.

Mendel's work was ignored in his lifetime but was rediscovered at the turn of the century, and quickly won wide acceptance. His 'factors' became known as 'genes' (a term introduced by Wilhelm Johanssen in 1903), and the variant forms of each factor became 'alleles'. Numerous traits that obeyed the Mendelian pattern of inheritance were discovered. This gave rise to the era of classical genetics in the 1920s and 1930s, closely associated with Thomas Hunt Morgan's experiments on the fruit-fly *Drosophila* at Columbia University. Classical genetics was based on the same technique that Mendel had employed, of crossing lines and studying the proportions of the different types of progeny. But it went beyond Mendel's work in one respect, namely the realization that his second law is not always valid: some genes tend to be inherited together, or are 'linked'. (The reason for this, we now know, is that the genes lie on the same chromosome.) Classical geneticists constructed detailed 'linkage maps', which quantified the extent to which the different genes in an organism tend to be co-transmitted.

An important discovery of classical genetics was that the relation between genes and traits is often more complicated than Mendel had envisaged. A single phenotypic trait may be affected by many genes, and a single gene may affect many traits. In the case of *Drosophila*, Morgan found that eye colour was affected by mutations in no less than twenty-five different genes. Nonetheless, Morgan argued that we can still sensibly say that a single mutant gene is 'the cause' of a particular fly's unusual eye colour, in the sense that the difference in eye colour between this fly and other flies is due to the former, but not the latter, carrying a mutated version of the gene in question. (Morgan claimed, controversially, that this was the standard meaning of the word 'cause' in science.) Today, Morgan's point is often expressed by saying that certain genes are 'difference makers' for phenotypic traits. That is, no gene gives rise to a trait all on its own, but genetic differences in respect of a single gene may explain phenotypic differences. This is essentially what medical researchers mean when they talk about a 'gene for breast cancer', for example.

Importantly, the gene of Mendelian and classical genetics was a theoretical entity, not something that was directly observed. Genes were introduced as hypothetical posits to explain the data from breeding experiments, much in the way that 19th-century physicists posited atoms to explain their data. To fulfil their explanatory role, genes had to be transmitted somehow from parents to their offspring, and to have a systematic effect on the traits that the offspring develop. But how exactly genes did this was unknown, nor was it known where genes were located nor what they were made of. Indeed, some classical geneticists did not believe that genes were real entities at all, just as some early physicists regarded atoms as useful fictions rather than real particles. In his Nobel lecture in 1933, Morgan noted that geneticists disagreed with each other on the reality of genes, but argued that this didn't matter. 'At the level at which genetic experiments lie', he wrote 'it does not make the slightest difference whether the gene is a hypothetical unit or...a material particle.'

In the eighty-five years since Morgan wrote this, the situation has changed dramatically. The gene has gone from being a hypothetical posit to something whose structure and function we know an immense amount about, in molecular detail, and which can be experimentally manipulated. Besides its intrinsic scientific interest, this change is philosophically interesting because it ties in with a perennial concern of the philosophy of science, namely to understand how scientific concepts evolve over time and how they 'latch on' to the world.

Molecular genetics

Molecular genetics came to fruition in the 1950s, the culmination of a long effort to understand the material basis of heredity. Since organisms develop from a single cell, the genetic material is presumably contained within the cell. But what is it made of? In the 1930s and 1940s, it was assumed that the answer is protein. For proteins were known to exhibit catalytic activity, that is, to speed up chemical reactions in cells; so if genes were proteins, this would explain how they are able to influence the traits of the developing organism. By contrast, nucleic acid such as DNA, a substance found in chromosomes, seemed like an unpromising candidate, since it is an inert and highly stable molecule. However, by 1950 an array of experimental evidence suggested that DNA is in fact the genetic material, which prompted researchers to try to determine its structure. This culminated in one of the most famous episodes in the history of biology: James Watson and Francis Crick's double helix model of DNA, published in 1953.

Watson and Crick's major discovery, building on previous work by Rosalind Franklin, was that DNA has exactly the right structure to serve as the genetic material. They showed that a DNA macromolecule is composed of two strands, entwined together in a helical structure (Figure 6). Each strand is formed by a long chain of repeating units called nucleotides, linked by covalent bonds. Each nucleotide has three sub-units, one of which is a nitrogen-containing compound called a base. There are four types of base: adenine (A), cytosine (C), guanine (G), and thymine (T). The two strands of a single DNA molecule are joined together by hydrogen bonds formed between their respective bases. Crucially, C is always paired with G, and A with T—a principle known as 'base pair complementarity'. This means that the sequence of bases on one strand of a DNA molecule, which can be represented as a long string of the four letters A, C, G, and T, determines the sequence on the complementary strand. Watson and Crick

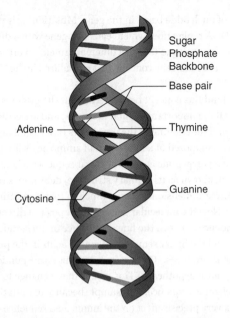

6. **DNA is composed of two strands, each made up of a long chain of nucleotides joined together. The two strands are connected by bonds between the nucleotide bases.**

realized that this suggests a possible mechanism by which the genetic material could be copied: the two strands uncoil, and each is used as a template for synthesizing a new strand. This is precisely what happens in DNA replication, and explains how genes are faithfully copied from cell to cell, and thus from parent to offspring.

The rise of molecular genetics led to a new scientific understanding of what genes are and how they work. The basic idea was that a gene is a particular segment of DNA located on a chromosome; that each gene gives rise to a specific gene product (a protein) in each cell, thereby influencing the organism's traits; and that which protein is produced depends on the precise

In the image, the following labels appear: Sugar Phosphate Backbone, Base pair, Adenine, Thymine, Cytosine, Guanine.

sequence of nucleotide bases in the gene. Mutations arise from errors in DNA replication, which leads to a gene with a different nucleotide sequence and thus to a different protein. In broad outline, this is the 'molecular gene concept' that crystallized in the 1960s.

To understand this concept better, recall that the proteins in a cell carry out all the important cellular activities, and are critical to the structure and functioning of the organism's tissues and organs. A protein is composed of a long chain of amino acids linked together in a polypeptide chain. The linear sequence of amino acids in a protein, known as its primary structure, determines the protein's three-dimensional shape, which in turn determines its behaviour. Now the molecular gene concept posits a direct correspondence between the linear sequence of nucleotide bases in a gene, and the linear sequence of amino acids in the protein that the gene produces. (Crick referred to this correspondence as the 'sequence hypothesis'.) This means that a change to the gene's nucleotide sequence, for example because of mutation, will have a very precise effect on the amino acid sequence of the resulting protein.

The molecular gene concept became established with the deciphering of the genetic code in the 1960s. To understand this, we need to briefly describe the process of gene expression, by which genes give rise to their protein products. This has two stages: transcription and translation. In transcription, a segment of DNA is copied to RNA, which is single-stranded. The resulting RNA strand is identical in sequence to one of the two DNA strands, except that thymine (T) is replaced with uracil (U). In eukaryotic organisms (which includes all plants and animals), the initial RNA strand undergoes processing to yield the mature messenger RNA (mRNA), which then leaves the cell's nucleus. In translation, the mRNA strand is decoded in a cellular 'factory' called a ribosome, where a growing polypeptide chain is formed by the addition of amino acids, one at a time.

Which amino acid is added to the chain is determined by the nucleotide sequence of the mRNA. Each triplet of nucleotides, or codon, maps onto one of the twenty different amino acids; thus CAG, for example, maps to the amino acid leucine. This mapping is the genetic code, and is (virtually) universal among living organisms. In essence, the genetic code means that the sequence of nucleotide bases in a gene can be regarded as a written instruction to build a protein with a particular amino acid sequence.

In the years after the genetic code was deciphered, molecular genetics progressed rapidly. The complexities of gene expression were unpacked, transforming biologists' understanding of how cells work, how organisms develop, and how genes exert their phenotypic effects. Moreover, thanks to remarkable technological advances, biologists were soon able to experimentally alter genes, for example by cutting and pasting DNA sequences, editing out parts of a sequence, and inserting DNA from one organism into another. Improved techniques for such 'genome editing' are still being developed today (such as the CRISPR technique that made the news headlines in 2015). With these advances, genes went from being passive objects of study to entities that can be manipulated at will.

In the 1980s and 1990s, the technology needed to sequence a gene, that is, to determine its precise nucleotide sequence, was developed. This culminated in the Human Genome Project, which published the complete sequence of all human genes in 2003. Since then the genomes of many other species have been sequenced too, and the field of genomics, which studies complete genomes and how they work, has progressed rapidly. Genomics has already led to striking new medical advances, particularly in oncology, and could potentially transform other areas too, such as agriculture. From its humble origins in Mendel's pea-plant experiments, genetics has travelled a long way.

Reduction?

How does the gene of Mendelian or classical genetics relate to the gene of molecular genetics? This question has long occupied philosophers of biology. One common-sense answer is that they are one and the same thing. That is, the hypothetical entity that Mendel and the classical geneticists posited to explain their data turned out to be a real entity, namely a segment of DNA that codes for a protein. On this view, then although there are two different gene *concepts*—Mendelian and molecular—these concepts pick out the very same object in the world.

This answer is not unreasonable. The idea that two concepts may refer to one object is standard in philosophy, as is the idea that a single scientific term may continue to refer to the same object despite considerable changes in associated scientific beliefs. For example, the term 'electron' as used by contemporary physicists and by late 19th-century physicists arguably refers to the same object, despite physical theory changing dramatically in the interim. So even though Mendel and the classical geneticists knew nothing of DNA or the genetic code, it is perfectly coherent to suggest that when they spoke of a 'factor' or 'gene', in the context of their breeding experiments, they were *in fact* referring to a segment of DNA that codes for a protein.

The suggestion that the Mendelian gene is identical to the molecular gene goes hand-in-hand with the idea that Mendelian genetics can be *reduced* to molecular genetics. Philosophers often speak of the reduction of one branch of science to another, meaning the explanation of the former's principles in terms of the latter. One example is the reduction of classical thermodynamics to statistical mechanics. Thermodynamics describes a closed physical system (such as a cylinder of gas compressed by a piston) in terms of its macroscopic properties such as temperature; while statistical mechanics describes the same system as an ensemble of microscopic particles in motion. The laws of thermodynamics can be derived from statistical mechanics, so in a sense the latter offers a deeper explanation of thermodynamic phenomena. This derivation is achieved with the help of *bridge principles* linking the vocabulary of the two theories, such as 'the temperature of a gas is the mean kinetic energy of its molecules'. Similarly, some philosophers have suggested that 'a gene is a segment of DNA coding for a protein' is a bridge principle linking Mendelian and molecular genetics.

At first blush, this suggestion is quite plausible. For classical genetics clearly got *something* right: the patterns of inheritance it discovered were real, and the explanation it offered of those patterns, in terms of genes that segregate in accordance with Mendel's laws, was broadly correct. But the explanation was fundamentally incomplete, given the lack of knowledge of what genes are, how they are transmitted, and how they affect an organism's development. These details were provided by molecular genetics, which offers a deeper explanation of hereditary phenomena, and to which Mendelian/classical genetics can therefore be reduced, the suggestion goes.

Despite its initial plausibility, this reductionist suggestion has been largely rejected by philosophers of biology. An early objection, due to Philip Kitcher, was that to explain why Mendel's law of segregation holds, we need only to note that genes lie on chromosomes which come in pairs, and that during meiosis (the cell-division process that produces haploid cells), the chromosomes in each pair separate; so each haploid gamete ends up with only one of each chromosome pair. Now this explanation is cytological or cell-level, not molecular-level. Adding in 'gory molecular details' does not improve the explanation, Kitcher argues, and so it is not really true that molecular genetics *explains* the law of segregation.

A second objection says that the alleged bridge principle linking Mendelian and molecular genetics is simply false, or at least it is an oversimplification, on the grounds that some segments of DNA count as Mendelian genes but not molecular genes, and vice-versa. One reason for this is that most of the DNA in any species' genome does not code for protein—a fact that was unknown to the originators of the molecular gene concept. Some non-coding DNA sequences play a key role in regulating gene expression, that is, in influencing which genes are transcribed and when. This allows a cell to produce a protein product as and when needed, and is the mechanism that underwrites cell differentiation, that is, the production of different cell and tissue types in a developing organism. Crucially, mutations in non-coding DNA, no less than in molecular genes, can affect the organism's traits. Now a non-coding DNA sequence will count as a Mendelian gene if it has a systematic phenotypic effect, such that variant forms of the sequence lead to variant forms of a trait. In short, encoding a protein is not the only way that a DNA sequence can play the role of a Mendelian gene.

A third objection is that some Mendelian genetic concepts resist definition in molecular terms. *Dominance* is an example. For any given dominant gene, there must be a molecular-level explanation for why the gene is dominant over its alleles—that is, for why organisms with one copy of the gene are phenotypically identical to those with two. The explanation may be complicated, but in principle it could be given. However, so far as we know, there is no single molecular feature shared by all and only dominant genes. That is, the class of dominant genes lacks any commonality that can be described in molecular terms. (In philosophers' jargon, dominance is 'multiply realized' at the molecular level.) Therefore, some hereditary patterns can only be captured by using Mendelian concepts.

A fourth objection argues that the relation between molecular and Mendelian genetics is more complicated than the reductionist picture suggests. The Mendelian/classical gene was associated with a distinctive experimental practice, and pattern of reasoning, which molecular genetics did not simply supplant. Indeed, the Mendelian gene concept is still alive today, which is not what we would expect on the reductionist account. For example, evolutionary biologists still use the Mendelian gene concept when they construct mathematical models of how genes spread in a population. In such models, a gene basically means any particle that is transmitted intact from parents to offspring, that obeys the law of segregation, and that has a systematic phenotypic effect. Similarly, the 'selfish gene' idea of Richard Dawkins, discussed in Chapter 4, is largely based on the Mendelian gene concept. As Dawkins notes, it makes no difference to the logic of his argument whether a 'selfish gene' is a gene in the molecular sense or not.

The consensus among philosophers that there is no simple reduction of Mendelian to molecular genetics may seem surprising. For molecular biology is often, and rightly, portrayed as a triumph of the reductionist method in science—that is, the method of studying larger systems by studying how their micro parts work. How can it be that the reductionist method has served molecular biology so well, and yet the Mendelian gene is irreducible to the molecular gene? In fact, there is no real contradiction here. For if a system, biological or physical, succumbs to reductionistic investigation, it does not follow that every macro-level pattern that the system exhibits can be eliminated in favour of a micro-level one. That molecular biology owes its success to the reductionist method is thus quite compatible with the concepts of Mendelian genetics still having an important role to play.

What is a gene?

Let us return to the question of what a gene is. Even if we set aside Mendelian genetics and focus just on molecular genetics, this question is still not straightforward. For since the 1970s, discoveries in molecular biology itself, ironically, have increasingly undermined the traditional molecular gene concept. Indeed, some authors speak of the 'dissolution' of that concept, on the grounds that it has turned out to be a rather imperfect approximation to reality.

Gene regulation, described above, is one source of the problem. The non-coding DNA sequences that regulate a gene's transcription are not generally adjacent to the protein-coding region, and may be located quite far from it; moreover, some non-coding sequences influence the transcription of many protein-coding regions. Biologists do not agree on whether non-coding sequences, including the promoter sequence that initiates transcription, count as part of a gene or not. If a gene is meant to include all of the DNA responsible for making a protein, as per the traditional molecular gene concept, then such sequences should be included. But including them makes the gene into a rather peculiar entity, with its parts scattered widely across the genome.

Still more problematic is the phenomenon of *alternative splicing*. In eukaryotes, the initial RNA transcript undergoes substantial editing to produce the mature mRNA that will be translated. 'Introns' are spliced out and the remaining 'exons' are then joined together. However, this splicing can occur in more than one way, which means that a single DNA sequence can actually give rise to many different proteins. Once considered a rarity, alternative splicing is now known to be ubiquitous. Organisms use it to their advantage, as it greatly increases the number of proteins that can be made from a single DNA sequence. But it casts doubt on the traditional molecular gene concept, because it undermines Crick's sequence hypothesis. The notion that a gene is a segment of DNA that codes for a single protein has turned out to be much too simple.

The picture is complicated still further by other findings. For example, translation can begin at different points on the messenger RNA, so a single mRNA transcript can produce a number of different proteins. Moreover, some protein products come from an RNA transcript that is itself derived from multiple DNA sequences, which may be located far apart in the genome. And in cases of 'overlapping genes', a single nucleotide sequence may be shared by (what are usually regarded as) two separate genes. Such complexities, of which there are many, raise difficult questions about exactly which DNA sequences should be counted as genes, and why.

For the most part, practising geneticists take a fairly relaxed attitude towards these questions. Their day-to-day research appears to proceed smoothly without the need to legislate on exactly what should count as a gene. From a philosophical perspective, this is itself interesting. For according to one popular view, scientific knowledge is distinguished from everyday knowledge by its greater precision. Now molecular genetics certainly exhibits a very high level of *experimental* precision. It is natural to think that this should go hand-in-hand with *conceptual* precision; that is, that the key scientific terms should be sharply defined. The case of genes shows that this is not always so.

Genes and information

The idea that genes contain 'information' has long been a defining feature of molecular genetics. It is often said that genetic information is transmitted from parents to offspring, and that it guides the development of an embryo into an adult. Familiar though this way of speaking is, on reflection it is actually quite puzzling. For most sciences do not use the language of 'information' to describe the causal processes that they study. Why then do geneticists employ such language to describe the processes of DNA replication and gene expression, and what does it mean?

In one valid sense of the word, information exists whenever we can predict one thing on the basis of another. The colour of the clouds contains information about the likely rainfall, for example, since the former predicts the latter. In this sense, genes certainly contain information about the phenotypic traits that they cause, and about the proteins they encode. For example, a gene that affects an organism's eye colour will thereby contain information about that trait. However, it seems unlikely that this is what biologists mean by genetic information. For in this sense, environmental features could equally be said to contain 'information' about phenotypic traits. The temperature at which a seed is incubated predicts the height of the adult plant; but biologists do not talk about 'environmental information'. So the rationale for talk of genetic information presumably lies elsewhere.

One popular suggestion locates this rationale in the fact that the genetic code is *arbitrary*. What this means is that the mapping between nucleotide triplets and amino acids, that determines which protein a given DNA sequence makes, is not chemically necessary. There is no chemical reason why CAC maps to histidine rather than leucine, for example. (This is why the genetic code had to be deciphered experimentally; it could not be read off the laws of chemistry.) The arbitrariness of the genetic code means that the DNA–protein relation is akin to the relation between a sign and its meaning. Consider, for example, a word in a natural language, or a traffic sign. Both carry information, in virtue of their conventionally assigned meanings. Since these conventions could have been different, the meanings are clearly arbitrary. Given that the genetic code is also arbitrary, this entitles us to regard a DNA sequence as containing information about, rather than merely causing, a protein's primary structure.

A different suggestion, due to John Maynard Smith, locates the rationale for talk of genetic information in facts about gene expression. The genes of all organisms contain regulatory sequences which act as 'switches', turning the gene on or off. They do this by binding to proteins known as transcription factors. These proteins are themselves coded for by other genes in the genome. Now in principle, any gene could be switched on by any transcription factor, if it evolved a suitable regulatory sequence. So there is an element of arbitrariness here too. Thus we can think of one gene as 'sending a signal' to another, telling it to switch on or off. The significance of this is that where there is signalling, it makes sense to talk of information. A honey bee's waggle dance is a signal whose function is to direct other bees to the nectar; thus the dance contains information about the nectar's location. Similarly, we can think of genes as transmitters and receivers of information, given that gene regulation relies on an evolved system of signals.

Against these suggestions, some authors argue that talk of genetic information is dispensable, as it plays no serious theoretical role in biology. (The philosophers Sahotra Sarkar and Paul Griffiths have described genetic information as a 'metaphor' masquerading as a precise scientific concept.) Others have gone further and argued that genetic information is actually a harmful notion, as it is bound up with an erroneous way of thinking about both development and inheritance. There has been a widespread tendency, past and present, to assume that the genome contains a complete set of instructions for building an organism, and that an organism's development is thus under exclusive genetic control. In Fox Keller's words, the genome has been seen as 'an executive suite of directorial instructions'. This view is not entirely wrong—genes are indeed crucial to development—but it is not the whole truth either, for two reasons. First, it turns out that genes themselves do not 'decide' when and where to make their protein products; rather, the cell regulates gene expression in response to environmental conditions. So the genome is a *reactive* entity, its behaviour partly under the control of the environment. Second, we now know that environmentally induced changes in a gene's state of expression, known as 'epigenetic marks', are sometimes transmitted from parents to offspring, and can affect the offspring's traits. So the linear DNA sequence itself is not the only inherited resource that organisms draw on in development.

For both of these reasons, it is oversimplified to regard the genome as containing a fixed set of instructions, or blueprint, for building an organism. Those who oppose talk of genetic information fear that it encourages this oversimplification. It remains to be seen whether future generations of biologists will heed their strictures.

Chapter 7
Human behaviour, mind, and culture

Our own species, *Homo sapiens*, is quite an unusual one. Our intelligence, language use, cultural inventions, technological prowess, and social institutions seem to set us apart from other species, even our fellow great apes. It is sometimes argued that this is an illusion of perspective, or the product of unjustified anthropocentrism, but this is a minority view. Certainly, we must be careful not to underestimate the sophistication of non-human animals, and it is true that many attributes once thought to be distinctively human, such as tool use, have been found in other species too. But even so, an unbiased observer from another planet would almost certainly single out *Homo sapiens* as distinctive. After all, no other species has been to the moon, built an opera house, invented parliamentary democracy, or destroyed the natural environment as we have.

Can biology shed any light on humanity and its achievements? One way to tackle this question is to ask whether human *behaviour* can be understood in biological terms—for it is our behaviour patterns that ultimately underpin our distinctive way of life. This latter question divides scholars into two broad camps. Those who answer 'yes' point out that *Homo sapiens* is an evolved species like any other, that the human mind is an evolved organ, and that human behaviour, no less than animal behaviour, has been shaped by natural selection. Those who answer 'no' accept that humans are evolved but argue that we have largely transcended our biological nature. How humans behave depends more on social norms and cultural expectations than on genes, they claim. So it falls to the social sciences such as anthropology and sociology, rather than biology, to explain human behaviour. Many traditional social scientists, unsurprisingly, have endorsed this viewpoint.

As with most intellectual divides, there is plenty of room in the middle ground here, despite what extremists on both sides have claimed. First, it is quite possible that biology can illuminate some aspects of human behaviour but not others; it need not be all or nothing. For example, our food preferences may well be biological in origin, but our participation in organized sport is probably not. Second, the availability of a biological explanation may depend on the 'grain' at which human behaviour is described. At a coarse grain, we may find behavioural commonalities that reflect our shared biological nature; but at a finer grain, we may find behavioural differences that are non-biological in origin. For example, pair-bonding is a universal human trait, but specific marriage customs vary from society to society. This means that biological and non-biological explanations of human behaviour will sometimes be complementary, as their explanatory foci will differ. Third, the terms of the debate may be questioned. Given that human behaviour is the result of multiple causal factors, some scholars regard the very dichotomy between biological and non-biological explanations as untenable. These three considerations mean that the space of possibilities is quite large, so we should be wary of anyone offering a simplistic answer to the question of whether biology can tell us anything useful about human behaviour.

Nature vs nurture

Most people are familiar with the nature vs nurture debate, which asks whether inherited or environmental factors 'make us what we are'. Scholars and others have long been fascinated with this issue, and it can arouse strong passions. One reason for this is that those on the 'nature' side of the debate, in particular, have often had rather thinly veiled political agendas. For example, the many Victorian scientists who held that 'feeble-minded' people (usually those with learning disabilities) had inferior genes were often keen eugenicists, advocating compulsory sterilization to prevent the deterioration of the gene pool. More recently, in 1994 the psychologists Charles Herrnstein and Richard Murray published *The Bell Curve*, in which they claimed that black people in the US score lower than white people on IQ tests because of their genes, rather than because of poverty or educational inequality. On the basis of this (highly contested) claim, Herrnstein and Murray went on to advocate ending government assistance for the poor, which they thought encourages low-IQ individuals to reproduce. Similar examples could easily be multiplied.

Can we address the nature–nurture question in purely scientific terms, that is, while divesting it of all social and political implications? This is itself a controversial issue. The 18th-century philosopher David Hume famously argued that one cannot deduce an 'ought' from an 'is'. According to Hume, statements about how the world *is* are one thing, while statements about how it *should be* are something completely different. Hume's dictum is still popular today; it underpins the widely held view that it is the job of scientists to discover the objective facts, and the job of policy-makers and legislators to decide what to do with those facts. Applied to the nature–nurture debate, this means that we should sharply separate the scientific question of whether a human trait is genetic in origin, from the political question of what, if anything, society should do with this information. For example, whether homosexuality is 'in the genes' has nothing at all to do with whether homosexuals should be accorded equal rights as heterosexuals, and similarly for other traits.

Sensible though this sounds, the idea that science is entirely value-free, concerned only with 'objective facts', is somewhat naïve, for two reasons. First, value judgements can be implicit in how scientists describe the world. The Victorian scientists who wanted to prevent the 'feeble-minded' from reproducing were not merely guilty of overstepping the bounds of science by straying into matters of public policy. They were also guilty of a deeper error, namely believing that 'feeble-mindedness' is a valid descriptor in the first place; and this belief reflected their ideological convictions. Second, scientists need to decide what to study, and value judgements often inform such decisions. Herrnstein and Murray wrote a whole book about racial differences in IQ scores, but nowhere do they explain why anyone should care about them. Similarly, recent evolutionary psychologists (see below) devote considerable effort to studying alleged differences between the male and female brain, but rarely do they explain why such differences, even if they exist, are of any importance. Thus the choice of which 'objective facts' to study may itself reflect individual or societal values.

These caveats are important, but it still makes sense to *try* to address the issue in purely objective terms. Let us then ask what modern biology teaches us about the traditional nature–nurture question? One moral is that in many cases, the question is not actually well-defined. This may seem surprising, since as we know biologists distinguish between genetic and environmental influences on an organism's phenotypic traits, which seems to correspond to the lay distinction between nature and nurture. Moreover, in biomedical science, certain diseases are often described as 'genetic', while in the study of animal behaviour, certain species-wide behaviours are often described as 'innate'; and both these terms appear to be synonyms for 'due to nature'. Why then is the nature–nurture distinction problematic?

One reason stems from developmental biology. For it turns out that genetic and environmental factors are almost always co-implicated in a trait's development, particularly for complex traits such as behaviours. As a result, even traits traditionally classified as 'genetic' may be susceptible to environmental modification. A classic example is the human disease phenylketonuria (PKU), which results from a mutation that affects the ability to metabolize the amino acid phenylalanine, leading to brain damage. However, if an infant with the mutation is kept on a diet low in phenylalanine, their brain will develop normally. So in fact, PKU is not purely 'genetic', for it is caused by a combination of a genetic factor (the mutation) and an environmental factor (consuming phenylalanine), neither of which does the damage alone. This means that the nature–nurture question is not well-defined *at the individual level*. That is, if we take a single individual with PKU, it makes no sense to ask whether their disease is due to their genes *or* to their consumption of phenylalanine: it is due to both. A similar moral applies quite generally, to traits of all sorts.

Importantly, the question may still make sense at the population level. That is, we can still ask whether the phenotypic *differences* in a population are genetic or environmental in origin. If all individuals in a population eat a diet high in phenylalanine, then the fact that some develop PKU while others do not will be due to genetic differences between them. And in fact, scientists who study the genetics of human behaviour have usually focused on the population level. Their main tool is 'heritability analysis'. This involves studying relatives in order to produce a numerical estimate of a trait's heritability, defined as the fraction of the trait's variation that is due to genetic variation. To illustrate the logic, suppose that two identical twins, separated at birth, are found to be much more similar, in respect of some trait, than two randomly chosen population members. Since the twins grew up in different environments, we can infer that their similarity is probably due to shared genes, that is, the trait has a high heritability. Many human behavioural and cognitive traits have been found to have moderate to high heritability. They include personality features such as extraversion and agreeableness; cognitive attributes such as IQ and musical ability; psychiatric conditions such as anxiety and schizophrenia; and social attitudes such as conservatism and religiosity. To many researchers, this suggests that the genetic influences on human behaviour are substantial.

However, the correct interpretation of heritability analysis is controversial. Its proponents regard it as a quantitatively precise way of addressing the traditional nature–nurture issue. This is sometimes justified, but there are a number of complications. First, a trait's heritability is inherently population-relative, and can vary substantially from population to population. Second, it is wrong to equate high heritability with 'genetic' and low heritability with 'environmental', contrary to what is sometimes thought. Take the human trait of having two legs. In a typical population, the only people without two legs have lost one or both in an accident—so differences in respect of this trait are not genetic. This means that two-leggedness has a heritability close to zero; but clearly, it would be wrong to infer that genes play no part in explaining why humans develop two legs. Third, where there is *gene–environment interaction*, the heritability statistic loses much of its meaning. 'Interaction', in this context, means that the causal effect of a gene on a trait is itself environment-dependent. For example, a given gene might increase the risk of anxiety in warm climates but decrease it in cold climates. In such cases, nature and nurture are inextricably entangled, even at the population level. Finally, *gene–environment correlation* occurs when genetically similar individuals are likely to experience similar environments. As with interaction, such correlation prevents heritability analysis from yielding a clean separation of genetic from environmental influences.

Let us summarize. At the individual level, all traits arise from a combination of genetic and environmental factors, and no clear meaning attaches to the question of which is more important. At the population level, the relative importance of genetic and environmental factors in explaining trait differences can sometimes be quantified, but not always. These points apply to all traits, including human behavioural and cognitive traits. So although biology may not resolve the age-old nature–nurture question, it certainly helps to clarify its meaning.

From sociobiology to evolutionary psychology

In the 1970s, the nascent discipline of sociobiology offered a bold new approach to the study of human behaviour. Pioneered by the Harvard biologist Edward Wilson, sociobiology applied evolutionary theory to the social behaviour of humans, and to the structure of human society more generally. The basic premise was that human behaviour is strongly influenced by genes and has evolved by natural selection; so Darwinian explanations can be given of particular behaviours and social arrangements. One of Wilson's best examples was the *incest taboo*. Though sexual mores differ widely among human societies, incest is forbidden in virtually all; and humans are instinctively averse to incest. Why so? Various anthropological explanations have been suggested, but Wilson argued that there is a simple Darwinian explanation. The offspring of incestuous couplings often have congenital birth defects, so there would have been a strong selection pressure against incest. This is why individuals evolved to be averse to incest and why societies prohibit it, Wilson argued. Another, more speculative example was the existence of male homosexuality. Invoking the theory of kin selection, Wilson argued that even though homosexual behaviour results in reduced fitness for the individual male, it may confer indirect fitness benefits on the male's relatives. Therefore, genes for homosexuality could be maintained in a population by natural selection.

Sociobiology was the focus of intense controversy in the 1970s, much of it intemperate, and often played out in the media. Critics saw it as a reactionary enterprise that would open the door to eugenics. In retrospect, it is clear that many of the critics were politically motivated and had mispresented Wilson's work. However, valid scientific criticisms were also raised, of which three deserve mention. First, Wilson's confident prediction that the social sciences would become a 'branch of biology' was indefensible. Adopting an evolutionary approach to human behaviour does not threaten to do social scientists out of a job. For social scientists are typically concerned with proximate rather than ultimate explanations (see Chapter 2). Second, sociobiologists tended to oversimplify the link between genes and behaviour, often talking as if genes rigidly determine specific behaviour patterns. This is untrue both because it ignores environmental factors, and because it neglects the role of cognition in producing human behaviour. Third, sociobiological explanations are most plausible when a behaviour is universal (such as incest avoidance). However much human behaviour is both highly *plastic*, that is, liable to change depending on the circumstance, and is variable across cultures.

These two last criticisms were taken on board by evolutionary psychology, a successor discipline to sociobiology that emerged in the 1980s and flourishes today. The main innovation of evolutionary psychology was to seek adaptive explanations not of human behaviour directly, but rather of its cognitive or psychological underpinnings. Its proponents argue that although human behaviour is variable and subject to cultural influence, there is nonetheless a universal psychology shared by all humans that evolved by natural selection, and that strongly constrains our behaviour. It consists of a set of 'mental modules', they argue, each of which performs a single specialized task. Actual behaviour results from the triggering of a module in a specific setting. Examples include a module for language processing, a module for recognizing others' faces, a module for mate choice, and a module for detecting 'cheats' in social exchanges. This modular picture, sometimes called the 'Swiss Army knife' model of the mind, contrasts with the traditional view that humans solve different tasks by using a single all-purpose psychological mechanism. Evolutionary psychologists argue that a modular organization is more efficient, as it allows adaptive behaviour to be more readily produced.

Though evolutionary psychology is adaptationist in spirit, its proponents do not think that the behaviour of modern humans is always adaptive. They argue that the human mind is well-adapted to the hunter-gatherer lifestyle that prevailed for most of human evolution. But in the last 12,000 years the environment has changed rapidly, and genetic evolution has not had time to catch up. So our 'stone-age minds' can lead to behaviours that are maladaptive in the modern world. This 'evolutionary mismatch' hypothesis, as it is known, is quite plausible in some cases. For example, our craving for sugar was likely adaptive in the environments where it evolved but leads to obesity in modern sugar-rich environments. More controversial examples include addictive behaviours, workplace stress, and postpartum depression, each of which has been argued to arise from evolved psychological tendencies that would have been adaptive in the Pleistocene epoch.

Though evolutionary psychology represents a methodological improvement over sociobiology, and has led to much interesting research, it is not free of controversy. In part, this is because some evolutionary psychologists have had a curious obsession with human sexual behaviour (including rape), and with male–female differences—topics that are inevitably sensitive. Also, critics have argued that evolutionary psychologists' intellectual commitments, such as their belief in a universal human psychology and in genetically hardwired mental modules, go beyond the available evidence. Another line of criticism accuses evolutionary psychology of the naive adaptationism critiqued by Gould and Lewontin (see Chapter 3), which assumes ahead of time that an adaptive explanation of any trait can always be found. Though these criticisms are not without substance, particularly in relation to popular works of evolutionary psychology, the best work in the field does meet the highest scientific standards.

A final criticism, which can in fact can be levelled against all theories that posit a genetic basis for human behaviour, is that they conflict with our sense of free-will. We humans intuitively believe that our actions result from our conscious choices, that is, that we are free agents. But the existence of strong genetic influences on our behaviour, or our underlying psychology, seems to threaten this belief. For example, a woman who is asked why she married her successful businessman husband will probably reply that she loves him. But how can this be squared with the evolutionary psychologists' claim that women have a hard-wired 'mate-choice module' that leads them to seek out high-status males as mates? Moreover, the practice of holding people responsible for their actions also seems threatened. If adult males have genes that make them aggressive, how can we blame a male for an act of road rage? Surely he can reply that 'his genes made him do it'?

Two responses can be given to this argument. First, no one seriously suggests that human behaviour is entirely genetically determined. At most, there may be genetic dispositions, of varying degrees of strength, to behave in particular ways. Second, and more importantly, what threatens our sense of free-will is really the idea that our behaviour is caused rather than freely chosen; the causes being *genetic* is not actually relevant. Environmental causes are equally threatening. Suppose it turns out that adult male aggression is caused by being corporally punished as a child. The man who commits road rage can still protest his innocence, claiming that his childhood experiences, rather than his genes, led him to do it. As this example shows, how to make room for free-will in a world of causes is a quite general philosophical problem that arises for everyone. It is not peculiar to sociobiology nor evolutionary psychology, so should not be levelled as an objection against these theories in particular.

Cultural evolution

A quite different way of applying Darwinian ideas to human behaviour is known as cultural evolution, or dual inheritance theory. It starts from the observation that there are striking cultural differences between human groups. Think, for example, of how family arrangements, funeral practices, and architectural styles differ around the globe. Since these cultural differences have emerged rapidly, in the space of a few thousand years, we can be sure that they do not stem from genetic differences. So human culture seems to 'float free' of underlying biology. Despite this, it is still possible that culture could evolve by a Darwinian-like process. For in humans, there are two inheritance channels that operate in parallel, genetic and cultural. Just as we inherit genes from our biological parents, so we inherit cultural practices and beliefs from our 'cultural parents'—who may be our biological parents or other members of our social group. This means that in principle, natural selection could operate on cultural as well as genetic differences, leading some cultural practices to spread through a population and others to decline.

Proponents of cultural evolution maintain that this has in fact happened. Think for example of how food provisioning has evolved over time. For most of our history, we were nomadic hunter-gatherers. This began to change some 10,000 years ago, when humans began to cultivate crops and domesticate animals, at a number of different locations around the world. Farming and agriculture then spread quickly, displacing the traditional hunter-gatherer lifestyle, and within a few millennia were found worldwide. This was partly the result of imitation, as hunter-gatherers saw the benefits of farming and switched, and partly the result of conquest and colonization. Now of course, scholars have long known that the spread of agriculture radically transformed humanity. But the distinctive claim of cultural evolution theory is that this was a genuinely Darwinian process, in which a superior cultural variant (farming) out-competed an inferior one (foraging), analogous to the way in which favourable genetic variants outcompete their alleles in ordinary biological evolution.

In the genetic case, the cumulative effect of natural selection is to adapt populations to the environment and to create diversity, as we saw in Chapter 2. Cultural evolutionists argue that the same is true when natural selection acts on cultural variation. To illustrate, consider the striking ability of humans to adapt to their local environment, that has allowed humans to become the most widespread and successful species on the planet. Think, for example, of the igloo-building skills of the Inuit, the hunting skills of the Kalahari bushmen, or the boat-building skills of the Vikings, all of which were crucial to survival in the respective environments. Now such skills are transmitted culturally, not genetically. There is no 'gene for building igloos' that the Inuit passed to their offspring; rather, the young were taught the skill, and it was refined gradually over many generations. The same is true of many other cultural practices. Thus cultural evolution is crucial to explaining how human populations adapt to their environment, and how the cultural differences among them emerge and persist.

How do cultural evolution and biological (or genetic) evolution relate? In one respect, the former depends on the latter, but in another respect they are autonomous. To see the dependence, note that cultural practices can only arise and spread because of humans' cognitive ability, which itself evolved by biological evolution. For example, both the initial invention of agriculture and its subsequent diffusion were only possible because humans were suitably intelligent, communicative, and had the ability to copy each other's behaviour. So if biological evolution had not led humans to evolve the requisite cognitive apparatus, cultural evolution could never have got going in the first place. To see the autonomy, note that cultural evolution does not depend on the presence of genetic variation, and occurs on a much faster timescale than biological evolution. Cultural variants can sweep across a human population far more quickly than genetic variants, since while genes are only transmitted vertically (from parents to offspring), culture can also be transmitted horizontally. Thus the speed with which a mutant gene can spread is constrained by the generation time, but the spread of a new cultural variant is not similarly constrained.

In some cases, there is an interesting interplay between cultural and biological evolution. The classic example is the spread of dairy farming. Domesticating cattle in order to consume their nutrient-rich milk began thousands of years ago, and spread by cultural evolution. However, the success of this cultural practice was limited by the fact that most humans lacked the ability to digest lactose. Once dairy farming had arisen, this then created a powerful selection pressure for a gene which would aid with lactose digestion. Eventually such a gene arose, and it spread by *biological* evolution through northern Europe and the Middle East, where it is found at high frequencies today. In contrast, the lactase gene did not spread to areas where dairy farming had not taken off, and it remains at low frequency in most human populations today, for example, in Asia. That is, we find a close correlation between a region's having a history of dairy farming and the presence of the gene for digesting lactose. The interesting point to note here is the interaction between biological and cultural evolution. The spread of a cultural practice—dairy farming—created the conditions needed for a process of biological evolution to alter the genetic make-up of the population to which the cultural practice had spread, which then allowed them to reap its full benefit. This interaction is known as 'gene–culture coevolution'.

One influential theory of cultural evolution was advanced by Richard Dawkins in *The Selfish Gene*. Dawkins described an evolutionary process that is unique to our species, in which rival 'memes' compete with each other for space in the human mind. A meme is meant to be a unit of cultural information, such as a song or a religious ritual, just as a gene is a unit of genetic information. Dawkins argued that like genes, memes are 'replicators', that is, entities of which copies are made.

Thanks to the human proclivity for imitation, memes leap from one person's mind to another, with occasional copying errors introduced. The spread of memes obeys broadly Darwinian principles, Dawkins argued. Memes that, for whatever reason, are better at getting copied will come to dominate the 'meme-pool'. Thus it is that we find pop songs that are catchy, and religions that exhort their followers to make converts; these are strategies that help the memes in question (the song and the religion) to spread. Dawkins envisaged a future science of memetics, parallel to genetics, that would study the principles by which memes spread in a population.

Though Dawkins's treatment was insightful, memetics has not come to fruition and most modern cultural evolutionists avoid the meme concept. In part, this is because of a standing ambiguity about exactly what counts as a meme, and how to count them. There seems no clear way to 'atomize' a complex cultural practice into distinct units. Does the Christian religion count as a single meme or many, for example? Moreover, Dawkins's meme concept was bound up with specific intellectual commitments that are not essential to cultural evolution theory, such as the controversial claim that memes have 'parasitized' the human mind for their own benefit. And, finally, the meme concept encouraged critics to dismiss the importance of cultural evolution prematurely, on the grounds that the analogy between genes and memes is fanciful and does not hold up in every respect. For these reasons, cultural evolution theory should not be tied to the meme concept.

One interesting philosophical issue is whether cultural and biological evolution should be thought of as similar in kind. Do they both exemplify the same abstract Darwinian logic? Cultural evolutionists argue that they do, but critics point to various dissimilarities. One is the point that culture can be transmitted both horizontally and vertically. Though true, arguably this does not make cultural evolution fundamentally unlike its biological counterpart. For although horizontal transmission of genes does not occur in humans, it is not biologically impossible, and indeed is very common in bacteria. A deeper difference is that novel cultural variants do not usually arise by chance, as novel genetic variants do, but rather are deliberate human inventions. A specific improvement to longboat design, for example, will likely have arisen not from a random accident, but from an intelligent Viking boat-builder realizing that the existing design could be bettered. A final, related difference is that cultural inheritance is 'Lamarckian', in the sense that modifications to a cultural trait that are made during an individual's lifetime can be transmitted. By contrast, biological inheritance is usually non-Lamarckian, since environmentally induced changes to an organism are not generally passed on to its offspring.

These last two differences are important, but their significance should not be overstated. The fact that a novel cultural variant is introduced deliberately, rather than arising at random, is quite compatible with its subsequent *spread* following a Darwinian pattern, that is, being due to the adaptive advantage that the variant confers on its users. Similarly, the fact that cultural inheritance is Lamarckian is compatible with treating the spread of culture as a Darwinian process. For one thing, Darwin himself believed in the inheritance of acquired characters; it was the neo-Darwinians who rejected Lamarckian inheritance. Moreover, recent discoveries in the field of epigenetics have shown that in certain cases, biological inheritance is itself Lamarckian, since acquired changes to a gene's 'epigenetic markers', rather than its DNA sequence, can be transmitted. For this reason too, we should not dismiss the idea that cultural evolution is a *bona fide* Darwinian process.

Finally, there is an interesting connection between cultural evolution and the levels-of-selection issue discussed in Chapter 5. We saw that while most biological traits evolve because of the advantage they confer on the individual organism, this is not always the case. Another possibility is that a trait evolves because it is group-beneficial, that is, by group-level selection. The same point holds true for cultural traits. An improved farming technique that spreads through a population will presumably benefit each individual farmer who employs it. But other cultural practices may evolve by group selection. For example, many traditional societies have complex systems of social norms that regulate their members' conduct. These norms are enforced by individuals punishing anyone who violates them. Now punishing a norm-violator is an altruistic action, since it is individually costly but group-beneficial. Social norms are thus good candidates for having evolved by cultural group selection.

In Chapter 5, we saw that group-level selection has often been regarded with suspicion in biology. But interestingly, the traditional objections to genetic group selection may not apply to its cultural analogue. In particular, one objection is that migration between groups will quickly render them genetically homogenous, thus diluting the between-group variation that is necessary for group selection to work. But in the cultural sphere no such problem arises. Migration between human groups does not automatically render them culturally homogenous, since migrants often adopt the cultural practices of their new group. Though the debate on this point continues, the indications are that group selection may be a more potent force in the cultural than in the genetic realm.

To summarize, while sociobiology and evolutionary psychology focus on the biological evolution of human behaviour and its cognitive underpinnings, cultural evolution theory takes a quite different approach. Based on the existence of cultural differences among humans, and the fact of gene–culture dual inheritance, it is suggested that a process of cultural evolution operates alongside genetic evolution in human populations, sometimes interacting with it. This theory has spawned fascinating empirical work, and raises interesting philosophical questions, such as whether cultural and biological evolution are fundamentally alike or not.

Conclusion

What can we conclude from this brief survey of the philosophy of biology? In addition to the specific morals drawn in each chapter, the most important general moral is that philosophical issues are pervasive in the biological sciences. This means that philosophical reflection on biology has a valuable role to play. By scrutinizing the meaning of biological concepts, studying the implications of biological theories, and probing the logic of biological explanations, philosophy helps to deepen our understanding of the worldview painted by modern biology.

Conclusion

Further reading

Chapter 1: Why philosophy of biology?

Good overviews of the philosophy of biology include *Sex and Death* by Kim Sterelny and Paul Griffiths (University of Chicago Press, 1999); *Philosophy of Biology*, by Alex Rosenberg and Daniel McShea (Routledge, 2008); and *Philosophy of Biology* by Peter Godfrey-Smith (Princeton University Press, 2016). Also useful are two collections of articles: *A Companion to Philosophy of Biology*, edited by Sahotra Sarkar and Anya Plutynski (Blackwell, 2008); and *The Cambridge Companion to the Philosophy of Biology*, edited by David L. Hull and Michael Ruse (Cambridge University Press, 2007).

Chapter 2: Evolution and natural selection

Darwin's argument is set out in *On the Origin of Species* (John Murray, 1859). Paley's design argument can be found in his *Natural Theology* (J. Faulder, 1802). A good discussion of Darwin and Paley is Francisco J. Ayala's 'Darwin's greatest discovery: design without designer', *Proceedings of the National Academy of Sciences* vol. 104, 2007. A good introduction to the neo-Darwinian theory is John Maynard Smith's *The Theory of Evolution* (Cambridge University Press, 1993). The logic of Darwinian explanation is explored by Elliott Sober in *The Nature of Selection* (University of Chicago Press, 1984), and by Daniel Dennett in *Darwin's Dangerous Idea* (Penguin, 1995). The proximate/ultimate distinction was set out by Ernst Mayr in 'Cause and effect in biology', *Science* vol. 134, 1961, and is critically re-assessed by Kevin N. Laland et al. in 'Cause and effect in biology revisited', *Science* vol. 334, 2011. The evidence in favour of evolution is set out in Jerry Coyne's *Why Evolution is True* (Oxford University Press, 2010). Elliott Sober's *Evidence and Evolution* (Cambridge University Press, 2008) is an advanced discussion of how evolutionary hypotheses can be tested against data.

Chapter 3: Function and adaptation

Good discussions of biological function include Philip Kitcher's 'Function and design' and Peter Godfrey-Smith's 'Functions: consensus without unity', both reprinted in D. Hull and M. Ruse (eds.) *Philosophy of Biology* (Oxford University Press, 1998). The aetiological theory is set out by Karen Neander in 'The teleological notion of "function"', *Australasian Journal of Philosophy* vol. 69, 1991. The causal role theory derives from Robert Cummins's article 'Functional analysis', *The Journal of Philosophy* vol. 72, 1975. The orthodox junk DNA viewpoint is challenged by Joseph Ecker et al. in 'Genomics: ENCODE explained', *Nature* vol. 489, 2012. W. Ford Doolittle replies in 'Is junk DNA bunk? A critique of ENCODE', *Proceedings of the National Academy of Sciences* vol. 110, 2013. Stephen Jay Gould and Richard Lewontin's critique of adaptationism is found in 'The spandrels of San Marco and the Panglossian paradigm', *Proceedings of the Royal Society B*, vol. 205, 1979. A special issue of the journal *Biology and Philosophy* 2009 contains papers re-assessing Gould and Lewontin's critique on the thirtieth anniversary of its publication.

Chapter 4: Levels of selection

Philosophical overviews of the levels of selection include Elizabeth Lloyd's article 'Units and levels of selection' in the online *Stanford Encyclopedia of Philosophy*, and Samir Okasha's *Evolution and the Levels of Selection* (Oxford University Press, 2006). George Williams's critique of group selection is found in his *Adaptation and Natural Selection* (Princeton University Press, 1966), and discussed by Elliott Sober in *The Nature of Selection* (University of Chicago Press, 1984). Hamilton's original papers on kin selection/ inclusive fitness are reprinted in his collection *Narrow Roads of Gene Land* vol. 1 (Oxford University Press, 1998). A recent philosophical discussion of Hamilton's work is Jonathan Birch's *The Philosophy of Social Evolution* (Oxford University Press,

2017). The kin versus group selection issue is examined by Elliott Sober and David Sloan Wilson in *Unto Others* (Oxford University Press, 1998), and by Samir Okasha in 'The relation between kin and group selection', *British Journal for the Philosophy of Science*, vol. 67, 2015. Dawkins's gene's eye view of evolution is set out in *The Selfish Gene* (Oxford University Press, 1976), and *The Extended Phenotype* (Oxford University Press, 1982). A good philosophical analysis of Dawkins's ideas is found in Kim Sterelny and Paul Griffith's *Sex and Death* (University of Chicago Press, 1999). The major transitions discussion stems from John Maynard Smith and Eörs Szathmáry's *The Major Transitions in Evolution* (Oxford University Press, 1995); good philosophical treatments include Peter Godfrey-Smith's *Darwinian Populations* (Oxford University Press, 2009), and the collection *The Major Transitions in Evolution Revisited*, edited by Kim Sterelny and Brett Calcott (MIT Press, 2011).

Chapter 5: Species and classification

Marc Ereshefsky's article 'Species', in the online *Stanford Encyclopedia of Philosophy*, offers a good overview of the species problem. Book-length treatments include John Wilkins's *Species* (University of California Press, 2009) and Robert Richards's *The Species Problem* (Cambridge University Press, 2010). Mayr's biological species concept is set out in his *Animal Species and Evolution* (Harvard University Press, 1963). A useful overview of species concepts is Jerry Coyne and H. Allen Orr's 'Speciation' in A. Rosenberg and R. Arp (eds.) *Philosophy of Biology: An Anthology* (Blackwell, 2009). The species-as-individuals thesis is set out by David Hull in 'A matter of individuality', *Philosophy of Science* 45, 1978; a good discussion is Thomas Reydon's 'Species are individuals, or are they?', *Philosophy of Science* 70, 2003. The widespread consensus that species do not have essences is challenged by Michael Devitt in 'Resurrecting biological essentialism', *Philosophy of Science* 75, 2008. The Linnaean classification system is discussed in Marc Ereshefsky's *The Poverty of the Linnaean Hierarchy* (Cambridge University Press, 2001). A useful introduction to phylogenetic systematics is found in David Hull's 'Contemporary systematic philosophies', in E. Sober (ed.) *Conceptual Issues in Evolutionary Biology* (MIT Press, 2008).

Chapter 6: Genes

Evelyn Fox Keller's *The Century of the Gene* (Harvard University Press, 2000) discusses genetics in historical perspective. Paul Griffiths' and Karola Stotz's *Genetics and Philosophy* (Cambridge University Press, 2013) offers a broad overview of philosophical issues in genetics. Philip Kitcher's article '1953 and all that', in *The Philosophical Review* 43, 1984, is the *locus classicus* for the view that Mendelian genetics cannot be reduced to molecular genetics. Alternative perspectives on reductionism are found in Sahotra Sarkar's *Genetics and Reductionism* (Cambridge University Press, 1998), and Ken Waters's 'Molecular genetics', in the online *Stanford Encyclopedia of Philosophy*. The gene concept is examined at length by Hans-Jörg Rheinberger and Staffan Muller-Wille in the article 'Gene' in the online *Stanford Encyclopedia of Philosophy*, and in their book *The Gene* (University of Chicago Press, 2018). The idea of genetic information is defended by John Maynard Smith in 'The concept of information in biology', *Philosophy of Science* 67, 2000; it is critiqued by Paul Griffiths in 'Genetic information: a metaphor in search of a theory', *Philosophy of Science* 68, 2001; and by Sahotra Sarkar in 'Decoding coding: information and DNA', in his *Molecular Models of Life* (MIT Press, 2004).

Chapter 7: Human behaviour, mind, and culture

A good overview of how evolutionary biology can be applied to the study of human behaviour is Kevin Laland and Gillian Brown's *Sense and Nonsense* (Oxford University Press, 2002). Kenneth Schaffner's book *Behaving: What's Genetic, What's Not, and Why Should We Care?* (Oxford University Press, 2016) offers a searching discussion of behaviour genetics, heritability analysis, and the challenges to the nature–nurture dichotomy. Edward O. Wilson's *On Human Nature* (Harvard University Press, 1978) is the original defence of human sociobiology; Wilson's ideas are critiqued by Philip Kitcher in *Vaulting Ambition* (MIT Press, 1985). John Tooby and Leda Cosmides outline evolutionary psychology in their 'The psychological foundations of culture', in H. Barkow, L. Cosmides, and J. Tooby (eds.) *The Adapted Mind* (Oxford University Press, 1992). Good philosophical discussions include Steve Downes's article 'Evolutionary psychology' in the online *Stanford Encyclopedia of*

Philosophy, and David Buller's book *Adapting Minds* (MIT Press, 2005). Cultural evolution theory is outlined by Peter Richerson and Robert Boyd in their book *Not by Genes Alone* (University of Chicago Press, 2005). Good philosophical discussions include Tim Lewens's book *Cultural Evolution* (Oxford University Press, 2015), and his article of the same name in the *Stanford Encyclopedia of Philosophy*. Cecilia Heyes' book *Cognitive Gadgets* (Cambridge University Press, 2018) integrates cultural evolution with aspects of evolutionary psychology.

Index

HIV/AIDS
A Very Short Introduction
Alan Whiteside

HIV/AIDS is without doubt the worst epidemic to hit humankind
since the Black Death. The first case was identified in 1981;
by 2004 it was estimated that about 40 million people were living
with the disease, and about 20 million had died. The news is
not all bleak though. There have been unprecedented
breakthroughs in understanding diseases and developing
drugs. Because the disease is so closely linked to sexual activity
and drug use, the need to understand and change behaviour
has caused us to reassess what it means to be human and how
we should operate in the globalising world. This *Very Short
Introduction* provides an introduction to the disease, tackling
the science, the international and local politics, the fascinating
demographics, and the devastating consequences of the
disease, and explores how we have — and must — respond.

'It won't make you an expert. But you'll know what you're talking
about and you'll have a better idea of all the work we still have to do
to wrestle this monster to the ground.'

Aids-free world website.

GLOBAL WARMING
A Very Short Introduction
Mark Maslin

Global warming is arguably the most critical and controversial issue facing the world in the twenty-first century. This *Very Short Introduction* provides a concise and accessible explanation of the key topics in the debate: looking at the predicted impact of climate change, exploring the political controversies of recent years, and explaining the proposed solutions. Fully updated for 2008, Mark Maslin's compelling account brings the reader right up to date, describing recent developments from US policy to the UK Climate Change Bill, and where we now stand with the Kyoto Protocol. He also includes a chapter on local solutions, reflecting the now widely held view that, to mitigate any impending disaster, governments as well as individuals must to act together.